ALIEN CONTACT:

PARADIGM SHIFT

by Derek Tyler

D0220870

The front cover logo is the unofficial unit patch of the 509th Bomber Group, which oversees the maintenance and operation of America's stealth aircraft. It was also charged with securing the crash site associated with the Roswell Event in 1947, as well as the recovery and transportation of everything that was found there, including survivors.

The phrase "To Serve Man" refers to an episode of "The Twilight Zone", a television series which aired during the mid-to-late 1960's. In it, a group of aliens landed on Earth, assuring everyone that they had come here only to serve us. They even had a book entitled "To Serve Man", which they claimed detailed how this was to be done. Eventually, a human was able to catch a glimpse of the book's contents. As the episode ends, you can hear her shouting "Oh my God! It's a...COOKBOOK!"

The phrase at the bottom of the patch, in Latin, translates exactly to "Tastes like chicken".

© 2018 by Derek Tyler.
ISBN-13: 978-0692948033
ISBN-10: 0692948031
Published by Burning Star Publications, LLC.

DEDICATIONS

I wish to express my eternal gratitude, sincere love and deep respect to the following individuals:

My mother and father, who allowed me to survive to maturity, even though they probably had far easier solutions available to them. Their boundless love and unceasing kindness has meant the world to me.

Best-selling author and radio host Debra Jayne East. Without her tireless encouragement and support, this manuscript would not exist. You can't have her—she's mine!

William White Crow (1954—2018), a man of great wisdom, supreme kindness & unwavering dedication to truth. His encouragement and willingness to share his knowledge with me changed my life. Though we had different parents, we were brothers all the same. We shall miss you, William.

Paul Sinclair, whom I consider to be one of the most dedicated, courageous and painstakingly accurate researchers in the world. He is also one of the finest men I know, and one of the best friends I will ever have.

Connie Willis (bluerocktalk.com), a long-time researcher, host of Blue Rock Talk on YouTube, and the weekend host of Coast-to-Coast Radio Network. Her generosity in inviting me onto her show as a guest gave the previous volume the traction it required, and the gift of her friendship is beyond price.

TABLE OF CONTENTS

CHAPTER 1: REFLECTIONS AND UPDATES

"I don't even bother about contradicting myself, because to me it seems a man who remains consistent his whole life must be an idiot. A growing person must contradict himself many times. Because who knows what tomorrow brings in? Tomorrow may cancel this day completely."
--Osho

"I watch the ripples change their size
But never leave the stream
Of warm impermanence"
-- David Bowie

I would like to begin with some comments about several things that appeared in the previous volume which I feel need some additional explanation. The first is an abduction memory I described, in which a pair of Grey aliens performed an operation on my feet. In describing the procedure, I stated that it seemed that my feet melted like candle wax, back to around my ankle bones.

This is, indeed, the memory I retained of the event. I do not, however, understand how something like this could be accomplished in practice, nor do I understand how my feet could be re shaped so quickly and seemingly effortlessly by the aliens afterward.

It is possible that my perception of what occurred is the result of a "screen memory", a false memory which was put in place when the original memory was erased from my mind. The argument against it being a screen memory is that the other memories I retained of that abduction experience are quite clear and do not appear to contain any gaps in logic or refer to anything which I cannot make sense of.

Even so, considering the difficulty of explaining how something like I described could be done in practical terms, it must be kept in mind as a possibility. Screen memories are certainly used with regularity during abductions, both to prevent abductees from being able to provide an accurate description of what took place and to add another layer of confusion about what happened within the mind of the abductee. Such confusion can, and often does, cause people to

1

assume that it was all nothing but a bad dream, and that an alien or military abduction did not in fact occur.

I suspect I will never understand, much less be able to explain, what it was that the Grey aliens did to my feet that night, or what the purpose of it may have been. I do not have any additional information I can add about it, only the memory as I have described it in the previous volume.

If my description was accurate (or even relatively close to being accurate), it would seem to involve things which are so far beyond my experience that I have no way of making sense of them. It is something I will surely always wonder about. It seems highly unlikely that my questions will ever be answered to my satisfaction.

Those who read the book may well find themselves in a similar position, and this is completely understandable. It was such a bizarre event that I would like to be able to make sense of it…but I can't. In this case, what you see is truly what you get: you know as much about it as I do, and your guess is every bit as good as mine. Should you decide that the most likely explanation is that it was a screen memory, I will not argue the point—that could very well be the correct explanation.

The next thing I would like to mention are a couple of statements which were made by the person I referred to as "The Colonel" in the first volume. As I mentioned in that chapter, I hesitate to disagree with or contradict his statements. As both a civilian and an outsider, I certainly would not do so lightly. It is his business to know about these things in detail, and he clearly has access to far more information than will ever be made available to me.

As I also stated in that chapter, on every occasion when I have taken it upon myself to disagree with The Colonel about something he said, later events have proven that he was correct, and that it was I who was mistaken. Considering all these things, I am naturally inclined to give him the benefit of the doubt concerning the statements he made to me.

Having said that, there are a couple of things The Colonel said which I find difficult to reconcile with the experience of myself and others. Though I am hesitant to disagree with him, it is also true that even those with an elevated level of access to the most carefully-guarded information in the government's possession are subject to being fed disinformation about various things.

People in such positions are far less likely to become whistle-blowers due to the much higher level of authority, power, salary and retirement benefits which come as perks of their jobs, but it is still possible that they could at some point opt to go public with their information. Should that eventuality come to pass, they must not be able to possess information which will allow them to blow the secrecy wide open all on their own.

To prevent this from happening they, too, are given certain information which is not factually accurate, as a means of hedging the government's bet. The specific things I find myself in disagreement with The Colonel about would seem to fit this requirement. They represent things he most likely would have had no way to discover were untrue on his own, but which at the same time conflict with the testimony and experience of other credible people who *would* be able to ascertain the validity of them. This is, I believe, the case with both issues I have with his statements, which I will now discuss.

At one point, The Colonel made the statement that "ALL" abductions of civilians are performed by the military, and that none of them were being performed by the aliens, whom he always referred to as either "the visitors" or "our guests". During my conversations with him, I do not recall him using the word "aliens" even a single time. I have heard this is also the case with certain other high-level sources, who refer to these beings as "the visitors" rather than as "aliens", though I am unable to remember now precisely who it was who reported this, or who it was the report referred to.

The Colonel stated that the military had various methods of making the abductions of civilians appear to have been the work of alien beings. I have no doubt that this is statement is accurate.

It would be quite logical and sensible, from the perspective of the military, to implement such procedures whenever possible. Causing military abductions appear to instead be abductions which were carried out by aliens greatly lessens the possibility that the military will ever be accused of the crime, much less held accountable for it. It adds an additional layer of deception to their actions, a buffer which serves to shield them and prevent them from being implicated or identified by those they choose to abduct.

3

In my opinion, however, it seems all but certain that alien beings are responsible for many abduction events as well. This is a viewpoint which is shared by virtually all the abductees I have spoken with, including those who recognize that they have also been abducted by the military.

I have no way of knowing just what percentage of the abductions the military may be involved in but, in my estimation, it is almost certainly not one hundred percent. I see of no reason to believe that aliens have suddenly decided to stop abducting humans, or that they no longer feel they have a need to.

This idea is supported by the fact that aliens have historically been reported to be abducting humans for many thousands of years, at a minimum, by cultures across the world which had no knowledge of, or contact with, each other at the time. Stories of gods from the skies taking human women for mates, and of human women bearing the children of these unearthly beings should not—and cannot reasonably—be interpreted as nothing more than ancient myths, without any foundation.

To make such a presumption is, it seems to me, to do our ancestors a great disservice and to drastically underestimate them. Those who lived thousands of years ago were no less intelligent or observant than we are today. They were people just like us, who lived without the advantages bestowed upon us by modern science, and the technologies which have resulted from it.

Our ancestors were neither savages nor fools. To treat their ancient reports as though they were the work of such is to make a grave mistake. It is not only untrue, it results in an unfair, inappropriate temptation to arbitrarily dismiss their statements in a wholesale manner. To do this is, it seems to me, both unwarranted and presumptuous. Simply because they may have lacked the ability to explain something, does not imply that nothing happened.

The people responsible for making, and recording, these reports did not do so out of some unspoken desire to attempt to fool those who came after them with invented tales of impossible events. In fact, doing such a thing would have been wasteful, pointless and counterproductive.

The reports of beings from the sky interacting and breeding with humans were written into the histories of these civilizations, right beside accounts of events which are known to have occurred just in the way they were described. There is no compelling reason to believe they were either fictional or intentionally inaccurate.

When those ancient stories are considered from the viewpoint of someone who is comparatively much more educated about the topic of alien abductions than our ancestors could have been, the connection becomes immediately clear. In fact, many of them are almost identical to modern-day sighting and abduction reports. To give our ancestors the credit they are due and presume that they were describing alien abductions in the best way they knew how, is the most reasonable—as well as the most likely—explanation for those records.

The idea that aliens are still likely to be abducting humans is also supported by the fact that it is known they were doing so prior to the original Greada Treaty. It is also known that they continued to do so— on a massive scale—after that treaty was established, in numbers which were in direct violation of the terms of the agreement.

There is good reason to believe that the abduction of humans has been a consistent activity, one which has taken place throughout the entire history of humanity on this world. To imagine, then, that they have chosen within the past half century to call a halt to the practice of abducting humans is, to my way of thinking, quite unlikely and is unsupported by any evidence, or any additional statements, that I am aware of.

Does the military abduct its own citizens, and try to make them believe it was aliens which were responsible for it? Absolutely. Is the military responsible for *all* the abductions of citizens, and the aliens blameless for them? I think it is not a realistic assessment of the situation. The Colonel's position was not one which would have involved his direct participation in abducting civilians for the military, and I think his statement that none of the abductions are being carried about by alien races was probably misinformed—an honest mistake on his part.

This leads directly to another of The Colonel's statements. He claimed that, although humans have lied to, cheated and broken promises they have made with the aliens, the aliens have never done so in return. This idea is contradicted by many sources, which claim that certain types of aliens have proven themselves to be inveterate liars. They are accused by various inside sources of engaging in dishonesty and deception constantly in their dealings with the American government.

One example of this involves the number of civilian abductions they carry out. It has been reliably reported by several sources that the (Type II) Greys abduct literally millions more people than the number they agreed to limit themselves to according to the Greada Treaty. Even though many abductions are clearly being carried out by the military, I see no reason to doubt that the Greys have lied about the number of people they abduct.

Another example can be found in the technology which has been turned over to us as part of the treaty agreement. It has been reported by multiple insiders that much of this technology is basically useless to us. Either it does not work as advertised, or it can only be made to work when certain additional parts—which are *not* made available to us—are installed or operated by the aliens themselves. As I understand it, this would be a direct, deliberate violation of the terms of the treaty. It would also technically be considered an act of war, as would the abduction of so many civilians as previously mentioned.

It has also been reported that the Greys have repeatedly lied to the American government about their origins and intentions. We were told, for instance, that they were humans from the future, who had traveled backward in time to attempt to correct a genetic problem they suffer from which originated during our time.

This is something which cannot reasonably be thought to be true. There is no reason to believe that humanity would naturally evolve into a form which is so much physically weaker and less robust than we are at present. Likewise, if we presume that humanity was intentionally modified by genetic engineering of our own design, there is no credible reason to believe that we would modify ourselves in such a way. The idea that Grey aliens are humans from the future, and that they have traveled backward in time seeking a way to prevent us from making a grievous error which resulted in their changed form, is yet another instance of deception on the part of the aliens.

There are others, as well. Even though literally dozens of different alien races are known to There is no reason I'm aware of to presume that alien beings either cannot or will not engage in dishonesty and

6

deception, when they feel it is to their advantage to do so. The fact that they are present on and around our world but choose to be there without publicly announcing their presence, demonstrates that deception is something they are very willing to make use of, when they think it is necessary.

It is also a clear indication that the primary reason certain alien races or groups have come here has nothing to do with our best interests. Rather, they are something which both the aliens and the government are concealing because these aliens pose a direct threat to us. There is no other logical and reasonable conclusion which can be drawn from the prolonged curtain of secrecy which enshrouds the entire affair, and the determined, unanimous denial of an alien presence by official sources.

If aliens in general had arrived to befriend and assist us, there would be little reason for almost a century of official denial on the part of the United States government. The fact that such denial has taken place anyway, and that it is such a high priority that the assassination of innocent American citizens to prevent the truth from being told is not only considered to be acceptable but has become standard operating procedure speaks for itself. The words it is saying are neither "friendly" nor "beneficial", and people who mistakenly think otherwise are not, in my opinion, thinking very clearly.

Do friendly aliens beings exist? Absolutely. Are they visiting Earth? Yes, some certainly are. But the official silence indicates that there are some very bad customers coming here as well, and that neither we, nor any alien friends we may have, are either willing or able to make them go away and leave us in peace.

I do not believe that The Colonel would have intentionally misled me about anything. He was not motivated to maintain the official program of secrecy or trick ufologists, but rather by his desire to make the public aware of certain information that he was unable to share with them himself, due to the nature and delicacy of his position. I could conceivably be wrong about both issues I mentioned here, but it seems unlikely, in my opinion.

I think it is most likely that The Colonel just made a couple of mistakes. History will eventually reveal the truth of the matter, and only then will it be possible to be completely certain about things. In the meantime, we try to be as careful as we can and to limit the number of mistakes we make as much as possible. It is the best we can do, and I think it is the most that can be asked or expected of us.

There is yet another interesting thing which has been made known to me by an inside source since the publication of the previous volume. It relates to my description of a human doctor holding a device which I stated was going to be used to literally cut my eyes out of my head. Since that volume was published, I have become aware of another abductee who reported the same thing, though that report has never been published.

It makes me sad that someone else has had reason to report such a thing. Although I of course realize that it is highly unlikely that anything was done to me which was not done to others as well, I had not heard of any other reports of such a thing happening during an abduction. I suppose I was quietly hoping that nobody else would have been forced to undergo such a horrible thing, that I would be the only one, even though I knew how unlikely that had to be.

I have been informed that my description of the event was inaccurate in an important way, and an explanation of why was provided to me. I accept the explanation, which came from someone who has direct knowledge of such things, and I will now share it with the reader. My goal, as always, is to provide the reader with the most accurate information possible, not to pretend that I never make mistakes.

According to this source, the eyes of the abductee are not actually removed, as I had presumed they were. "This is a trick they use on abductees," my source told me. "Alien medical technology, though it is far more advanced than our own, is not capable of repairing the damage that would be caused by the removal of the eyes and healing it up in such a way that it would appear not to have occurred.

"What happens is, the patient is shown the device—which you described very accurately, by the way—to make them *think* they are about to have their eyes cut out of their head. This has the effect of forcing the patient's body to produce a massive amount of adrenaline, releasing it into the bloodstream and sending it coursing throughout their entire body.

8

"At that point, a tube or two of blood is extracted from the victim, blood which has now been saturated with adrenaline. These tubes of blood are given to the Grey aliens, who swirl the blood around in their mouths, absorbing that adrenaline into their bodies. It gets them high, and they appear to love the rush they get from it. They want to get that feeling as often as they can.

"So, although the victim is caused to *believe* that their eyes are about to be removed from their sockets in an extremely brutal, horrifying and painful manner, it is not really done. The memory will be deleted, and they will eventually be returned to their homes.

"If, like you, they happen to retain any memory of the event, it will be the memory which is strongest—the memory of being shown that device and being given every reason to think it was about to be used to remove their eyes. Almost nobody will even remember that.

"But, you must keep in mind, to the people who do this, it takes on the feeling of working on an assembly line. They perform the same repetitive actions, time after time. Eventually, they become bored with it. They can lose their focus and fail to completely erase the memories. This is typically where abduction memories come from—they are the result of a mistake made by the medical team."

So: members of our own military black ops forces are being used to terrify abductees for the sole purpose of causing their blood to be saturated with adrenaline, and the blood is then extracted and used to get the Grey aliens "high". This is an explanation which fits in well with what I have heard from other sources, and it is one which makes sense to me.

This should not be construed as the *only* reason the military abducts its own citizens, but rather as something which is done in addition to other things during the time that an abductee is under their control. If it were the only reason for abductions, then any random group of people would suffice for the purpose—it would make no difference whom the adrenaline-saturated blood was extracted from.

The victims of abduction, however, are not chosen at random from the population, they are carefully selected individuals and they are often abducted repeatedly over time. This makes it clear that the military has other reasons for abducting its own citizens, and specific criteria they use to decide who will be targeted.

CHAPTER 2: REASONABLE EXPECTATIONS

Look up at the stars and not down at your feet. Try to make sense of what you see and wonder about what makes the universe exist. Be curious.
--Stephen Hawking

"How I wish, how I wish you were here
We're just two lost souls, swimming in a fish bowl
Year after year
Running over the same old ground
What have we found? The same old fears.
Wish you were here."
--Pink Floyd

Near the beginning of the previous volume ("Alien Contact: The Difficult Truth") I stated that, when one becomes involved with the study of alien contact, there is never an end to the learning. There will never come a time when any of us will have all the answers, and we will never reach a point where our knowledge of the subject is so vast and so complete that it becomes unnecessary for us to learn any more.

The path to knowledge in this arena is long, difficult and confusing. It is also strewn with an endless number of traps and dead ends, which are placed there intentionally and are ever ready to ensnare us or lead us astray without warning. From complete beginners to veteran researchers, and even to some of the most well-placed sources imaginable, a multitude of traps have been prepared with them in mind and disinformation has been purposely directed at them by those in charge of such things.

Nobody is immune to it and nobody is given a free pass. If you make the decision to study the topic of alien contact in a serious manner, there *will* be occasions when you believe things which turn out to be untrue. You *will* be victimized more than once by the vast, multi-layered array of disinformation which has been laid out with the intention of trapping people just like you, much the way a spider's web traps a fly.

There is no shame or dishonor in this, nor is it an indication that you have been careless or are prone to being gullible. It is an inherent part of ufology, and it is overseen and practiced by teams of highly-trained, highly-motivated professionals who are extremely good at what they do. Regardless of how careful you may be and how many precautions you may take against these things, you may be certain that there are more traps lying in wait for you than you will be able to find and identify. This is the reality of life as a ufologist, and if you spend enough time living it, it will eventually make its presence known.

The inherent complexity of the study of alien contact is tremendous. We are attempting, firstly, to understand as much as possible about beings who are unfamiliar to and very different from us. This is a difficult proposition even under the best of circumstances—and, as anyone who studies alien contact quickly realizes, "the best of circumstances" is a situation which will never apply to us as ufologists. Far from it!

The problem of understanding the nature of a single race of intelligent, non-human beings, as well as discovering their motivations for coming here, their agenda and their intentions regarding both Earth and humanity, would be daunting enough all on its own. The possibilities allow for a virtually unlimited number of scenarios, any one of which might apply, as well as for the possibility that the true scenario might very well be something which the aliens do not wish us to become aware of and therefore make every effort to conceal from us.

If this turns out to be the case, as it apparently has, we are already faced with a tremendous challenge as researchers. Reliable information is always difficult to come by, and if the aliens themselves are taking measures to prevent us from learning about them, the difficulty of doing so will be made far greater than it already was.

Add to this the fact that we are dealing here not with only a single alien race, but with what certainly appears to be a veritable multitude of them, all at the same time. This causes the difficulty of correctly understanding alien contact to be multiplied many times over.

Next, we must factor in the challenge of somehow coming to grips with the fact that all these non-human entities possess scientific expertise and technological capabilities which are far more advanced than our own. The gap between their level of capability and our own is far greater than anything we have previously experienced or are accustomed to dealing with.

Possibly the most accurate way for us to picture it is to imagine the difference between a primitive Stone Age tribe which is discovered deep in the jungle and our own modern technology. It is impossible to overstate the difficulties that members of such a tribe will experience when faced with the proposition of understanding things like video cameras, combustion engines and other devices which we consider to be so commonplace that we take them for granted.

When we add to this the idea of having a Stone Age people arrive at a correct understanding of things such as fighter aircraft, computers, satellites, alternating current, nuclear reactors, aircraft carriers, molecular biology, radio telescopes and the rest of the things which can be found within our society, the prospects are multiplied exponentially. Far before coming to a correct understanding of our technological capabilities, the members of the Stone Age tribe will surely feel their poor brains begin to melt and slowly drip from their ears onto the ground. It is simply a bridge too far, it is a leap they have no way to make and a gap they will be completely unable to cross.

We find ourselves in a similar position regarding alien technology. The aliens in question have varying levels of scientific knowledge and technical capabilities. Their civilizations have arisen at various times, and each of them have different innate levels of intelligence as well as different goals and agendas they have pursued.

There is no question that there have been numerous examples of technologies and information which have been shared between civilizations because of trade or diplomacy. All of them, however, are highly advanced when compared to our own level of technological advancement.

The gap between human civilization and their own probably has a considerably wide range. It is believed by the American military that there is a difference of at least 200,000-300,000 years between the capabilities of even the least-advanced of these civilizations and our own. At the other end of the scale, we are very likely looking at a difference of literally millions of years between their level of technology and our own, and it is entirely possible that some of these alien civilizations have been in existence long enough to push that gap out to tens, or even hundreds of millions of years.

It goes without saying that a technology gap of this immense size is not something which it is possible to close. Regardless of how much effort we may put into it, or how long we may spend trying, we are never going to equal or surpass the capabilities of these alien civilizations. It is pointless to pretend otherwise.

If an alien civilization has technical capabilities which are, say, ten million years more advanced than our own, we could literally spend a million years working to advance our level of technological know-how and still be looking at a gap of nine million years between their capabilities and our own. Even that would be a best-case scenario, one which would require them to have made no further scientific discoveries or advancements at all during that million years. We are operating from a position of tremendous—and permanent—disadvantage when compared to even the least-accomplished of these alien races, and we have no realistic way to close the gap on our own.

This comes as no surprise, of course, to any of you who are reading this book. It should serve as a constant reminder, however, of the intrinsic danger involved with any form of contact between ourselves and any of these alien civilizations. Even the least advanced of them certainly could put a quick end to us, if it wanted to badly enough.

We, on the other hand, would appear to have very little, if any, leverage we can use against them. Anything we have that they might happen to want, they surely can simply take, with our permission or without it, and there would be little we could do to stop them from doing so.

In military terms, the scenarios in terms of active, armed conflicts between us and any of them would appear to fall anywhere within the range of exceedingly grim to hopelessly impossible. One does not, after all, voluntarily declare war on a civilization which had already established trade routes between the stars at a time when one's ancestors were still nothing more than primitive apes living in the jungle and sleeping in trees at night. Unless, of course, they enjoy the prospect of being awakened during the night to the sound of anti-gravity bombs being detonated in their yard. I dislike making assumptions, but in this case, I think it is probably safe to assume that the prospect of having anti-gravity bombs detonated in your front yard is something we can all agree would make for a distinctly unpleasant experience.

Getting back to our original topic, the following statement has been made with respect to alien technology, and I fully agree with it: "Alien technology is not only more advanced than we imagine, it is more advanced than we *can* imagine." Just as the members of a Stone Age tribe would be utterly unable to imagine the technologies and capabilities possessed by a modern, industrialized society, we are similarly unable to truly imagine the potential capabilities of advanced alien civilizations.

Even attempting to guess about such a thing is for the most part an exercise in futility. The level of their scientific expertise and its resultant practical capabilities are surely well beyond our ability to predict or guess. Now the difficulty we are faced with when trying to arrive at a reasonably accurate understanding of the non-human beings which are present in and around our world, already formidable, is once again increased exponentially.

As if our job wasn't already more than problematic enough, our journey to knowledge will be intentionally made as difficult as possible by teams of highly-skilled professionals employed by our own government. It will not only place obstacles in our path at every possible point, it will do its best to ensure that we are not taken seriously by members of the public. It is hard to overestimate the number of problems which can be thrust upon us by a government which chooses to use every method at its disposal to make our investigation more difficult and our lives more miserable.

It is often the case that disinformation professionals practice their craft by simply encouraging people who have made a mistake of some kind to continue to make it, and if possible to turn what may initially be a minor misstep into an all-out exercise in futility which ultimately encourages others to be drawn into the same set of mistaken ideas.

Ideally, from the point of view of the disinformation professionals, it can eventually take on a life of its own—a life which they will carefully feed and nurture along the way—and result in large numbers of people falling prey to the same con. When this happens, they will have a natural tendency to form groups of like-minded individuals, whose members will then proceed to speak with great confidence about things which are not in fact true, encouraging each other's belief in those things in what effectively becomes a self-reinforcing echo chamber. It can be thought of as a snake which is attempting to swallow its own tail—the more of it the snake manages to swallow, the worse off its condition becomes.

Finally, we are forced to deal with the fact that our world is filled with a variety of frauds and con artists, and that this will likely always be the case. Despite the seriousness of the topic, there have been and are people who will attempt to exploit it for their personal gain. They invent stories about fictitious alien encounters or contact, and then market them to as many people as possible. They have been quite successful at doing this over the years, and many of these people have become multi-millionaires in the process.

I often say to people "If you want to find the truth, follow the trail of dead bodies." If someone is speaking the truth about alien contact, they will not be allowed to earn millions of dollars doing so, nor will they be allowed to attract a fan base of millions of people who breathlessly hang onto their every word. They would be selectively eliminated long before something like that could happen.

Those charged with keeping the secrets about alien contact are very serious and extremely lethal. Having someone attain worldly success by spreading the truth about alien contact is what they are *not* going to allow to happen. If, therefore, someone has become a multi-millionaire by writing or speaking about alien contact, you may be certain that they are *not* speaking the truth. When one considers all these difficulties which stand in our way, including the inevitable complexity of the aliens themselves, it becomes obvious that we cannot realistically hope to ever get to a point where we manage to get *everything* right.

The good news is that it is not necessary for us to be able to get absolutely everything right. If we can manage to get enough right, it should be adequate to allowing us to gain a reasonably accurate view of the big picture. Some things will surely always remain a mystery to us. Some other things, we may well find reason to change our minds about later. We will do the best we can, and trust that it will, in the end, be good enough to make the trouble worthwhile.

CHAPTER 3: DARWIN'S LAST STAND

"Everything you think you know is a lie."
-- Lloyd Pye

Oh, no! There goes Tokyo!
Go, go Godzilla!
--Blue Oyster Cult

Millions of years ago in Africa, we are told, apes came down from the trees and began to make their home on the ground. Over vast eons of time, they slowly evolved into a series of more advanced forms, many of which are to be found within the fossil record.

As time went on and Darwinian evolution continued to occur, these new hominid species became better adapted to their environment, as well as becoming increasingly intelligent as they continued to evolve and change. The more primitive proto-human forms eventually were replaced by newer, more capable evolutionary products which were able to out-think them, as well as being able to compete more successfully for food.

This resulted in the older, more primitive species to eventually die out as those which were better able to survive and thrive in the natural environment gradually became dominant. It was, we are told, a classic example of the "survival of the fittest", a basic component of Darwinian evolution. Over time, as the process continued, and the primitive simians became steadily more intelligent, more capable and more developed, the eventual result of this evolutionary chain were homo sapiens, modern humans as we know them.

There came a time, somewhere between 60-100,000 years ago, we are told, that early humans began to migrate northward. They left the continent of Africa, to try their luck in other lands they had discovered, which were yet uninhabited by man. They spread into the Middle East, and from there to the Indian sub-continent, and successfully established themselves there by building what are now considered to be the most ancient cities and cultures in the world.

As time passed and the need for additional land to colonize occurred, these early cultures dispatched explorers, followed shortly thereafter by groups of people who acted as colonists, in all directions.

The human population spread into what is now China, and from there into the rest of Asia. It also migrated north and east, adapting itself to the local conditions and slowly populating what is now Europe and Russia.

Ever-eager to explore and conquer new lands, bands of human settlers eventually made their way into the frozen north. They were responsible for creating new cultures, new kingdoms and new nations all along the way. North into what is now Scandinavia they expanded, and into Mongolia, the Himalayan mountain range and Siberia as well.

From Siberia, we are told, humanity eventually migrated east, across a narrow land bridge which linked Asia and what is now Alaska. Over time, they spread south and east, populating both North and South America and establishing great civilizations such as the Maya, Inca and Aztec empires.

This, in a nutshell, is the story science uses to describe the evolution of homo sapiens, its eventual expansion across and subsequent colonization of the world's major land masses. The scientific community has been virtually unanimous in accepting it as an accurate depiction of events for, at the time of this writing, the better part of two centuries.

The proposition that the combination of vast stretches of time and a series of genetic mutations resulted in the evolution of primitive apes into modern man was quickly adopted by educational systems throughout the industrialized world. They wasted no time in incorporating the idea into their curriculums.

Anyone who dared to disagree with this idea was—and still is—immediately attacked, discredited and vilified by the entire community of "serious" scientists. Though often disagreeing with each other regarding various matters, they have proven quite willing to lend their assistance when the time comes to close ranks and form a lynch mob.

Due to this quite natural presumption on the part of schools that evolution provides the only legitimate scientific explanation available, it is not considered necessary to put a lot of time and effort into scrutinizing it in a detailed manner within their classrooms. They see little to be gained by forcing students to memorize the Latin names of obscure, alleged human ancestors such as *Australopithecus afarensis, Homo habilis, Paranthropus robustus* and *Homo heidelbergensis* when they could be doing something useful, such as dissecting frogs, playing around with Bunsen burners or learning to write secret messages with lemon juice instead.

It is considered even less necessary to waste precious class time discussing its sole competitor, Creationism. After all, it does not take a genius to realize that, when only two possibilities exist and one of them consists entirely of unscientific nonsense, the other one *must* be correct.

Why are we discussing this topic here, in a manuscript which is intended to focus not on human evolution, but on alien contact? As with many things which may not at first appear to be related to the topic of alien contact, closer examination reveals that a connection does indeed appear to exist.

It is a connection that, if we do not lay the foundation for it now, a great many people will fail to consider. This is inevitable, due to both the programming we have received over the years and some rather important things that the schools made the decision to either downplay or not mention at all.

Discussing these matters now, early on, will pay dividends later by allowing us a more accurate and complete understanding of the Big Picture regarding alien contact than would otherwise be possible. Our schools and universities may have neglected to properly emphasize certain aspects of this topic, but I will not make the same mistake.

I understand that those who read this manuscript will be anxious to get to the sections which deal ideas that seem to be more directly related to alien contact, and we will get to them soon. First, however, we've got some other important business to take care of. It is well worth doing, and we are going to take care of it right now.

You see, there are several issues which arise when the classical theory of human evolution is inspected closely. These issues are far from being trivial. They are so problematic that, once they are properly taken into consideration, the story of human evolution we are taught to believe is, for all practical purposes, a settled matter turns out to be anything but that.

19

As we are about learn, the all but universally-endorsed mantra which describes a logical, orderly, well-documented evolutionary progression of primitive apes into modern man becomes increasingly difficult to accept the deeper one scrutinizes it. When subjected to methodical, dispassionate analysis, the picture of human evolution which is by now accepted as representing self-evident, scientifically-established truth at all levels of society is revealed to be deeply flawed. The problems inherent to it are of such a profound nature, in fact, that once they are properly factored in they give the whole story the distinct appearance of being only incorrect, but flat-out impossible.

Surprise. Disbelief. Shock! Horror! How on earth could I possibly make such a statement with a straight face, much less expect people to take the idea—or, at this point, me—seriously?

I will admit, accomplishing that is going to be quite a trick, quite a trick indeed. Stick around--I am about to show you how it's done.

There will be thrills, there will be chills! There will be danger, intrigue and all the suspense you could wish for along the way!

In fact, I'm going to do even better than that!

Listen. I know that you are probably already a little bit mad at me for making you read a chapter about evolution in a book that's supposed to be about aliens. I want to try to redeem myself for that, and I also want this chapter to be fun for you to read. * To that end, I am willing to push the stakes even higher, knowing even as I do that it will force me to increase the level of my game, if I hope to be able to keep up.

I am not afraid! I accept the challenge! Here, then, is my sneaky, underhanded plan, which my opponent will never see coming! I shall attempt to infuriate and rattle my opponent, by making use of a technique I learned long ago, in the days of my youth, by watching Gary Payton of the Seattle Supersonics basketball team. Gary was not only a monster on defense, he was one of the sport's legendary trash talkers. I'll give it a try and hope for the best!

* After long hours of careful deliberation, I decided this was probably a better idea than the alternative: making it an exercise in cruel, unbearable torture which would swiftly bring my career as an author to a humiliating, almost certainly painful and richly-deserved end.

"Yo! Listen up, evolution—if that is, in fact, your REAL name! By the time I'm done with you, your reputation is gonna remind people of Godzilla stomping over a carton of eggs as, like an unstoppable juggernaut of power and destruction, he approaches the doomed city of Tokyo!"

Please, people! Remain calm! Return to your seats! Violence is not the answer to anything! Don't hate the player, hate the game!

One of the key assumptions which is required for the standard theory of human evolution to give the appearance of making logical sense, is the explanation of how the Caucasian race came into existence. After migrating far northward out of Africa, we are told, humans populated what is now northern Europe. There, not far from the Arctic Circle, the sun's rays strike the Earth at a much shallower angle, compared to the direct rays which occur in the tropic lands— specifically, in Africa.

Due to the shallow angle of the sun's rays, we are told, the people who became the inhabitants of what are today the Scandinavian countries underwent several rather radical changes. The curly, black hair which they brought with them from Africa eventually lost its dark color and turned blonde, or sometimes bright red. Their eyes, once uniformly brown in color, changed to blue or green. Their dark-colored skin, now exposed to only indirect rays from the sun, lost its natural hue and, as time passed, became very pale. This was an example of natural evolution, we are told, and something which makes perfect sense, considering the change in climate.

Now, I have heard a lot of dumb ideas in my life, which makes it difficult to be certain whether this qualifies as the single dumbest idea I have ever heard—if not, it certainly must be close. It is also completely unsupported by any type of evidence—something which, as we will see, turns out to be a recurring theme when discussing evolution.

For evolution to occur, a change must in some way confer a survival advantage to the organism. If it is to eventually become the dominant characteristic of a species and replace the former—now somehow inferior—form, it must give the individuals who possess it a greater opportunity to survive and—most importantly—reproduce, than the previous version was equipped with. Without such a clear and evident advantage, there is no reason for a mutation such as blonde hair, blue eyes or pale skin to eventually predominate throughout an entire population.

How, I wonder, does blonde hair and blue eyes confer an advantage to an individual in terms of the ability to successfully reproduce? How does pale skin cause someone to have the ability to successfully bear and raise children, while their neighbors with darker skin find themselves unable to do so?

There is no reason I know of to think that such would be the case. But, even if it were, how do these physical peculiarities confer such an incredible, one-sided advantage to those who are born with them that all those who lacked them would eventually disappear completely?

What would cause them to dominate so completely that there would be no evidence at all that so much as a single person with dark skin and kinky black hair had *ever* existed in that land? Because that is exactly what we are expected to believe took place—there are no fossils of them to be found in Northern Europe, no stories told of them, no mention of them recorded in their histories. The people of Scandinavia, in fact, had absolutely no idea that such a race of people had ever existed.

Such overwhelming genetic dominance is, I am sure you will agree, quite a fortuitous advantage to have in one's possession. Though I am somewhat reluctant to introduce logic into what appears to be nothing more than a nonsensical fairy tale, I think we need to ask ourselves a few sensible questions and supply them with sensible answers before going any further:

If one were to drop off a group of African natives in Sweden, and an equal number of blonde, blue-eyed Caucasians among them, do we really believe that the Africans would find themselves unable to bear children?

Do we honestly believe that Caucasian children would be more able to survive to adulthood and, in their turn, reproduce, than children who happened to have a darker skin tone?

Is there any reason to realistically believe that pale skin and blonde hair would confer such a tremendous advantage to people that all those who lacked them would eventually, within the local population, become completely extinct?

How extinct does a race of people have to be, before all fossil evidence of them is obliterated and all memory that they had ever existed at all is lost?

It seems to me that the answer to the first three of those questions must surely be a resounding "No!", and that the answer to the fourth question must be "Very, *very* extinct!" According to the standard model of human natural evolution, however, the answer to the first three questions apparently must be "yes". Maybe you can think of a way to make sense out of answering "yes" to those questions, but I am unable to do so. As best I can tell, the explanation of how pale skin, blonde hair and blue eyes developed within the human population comes to a grinding halt right there.

Let's presume, for the moment, that the idea that pale skin, blonde hair and blue eyes appeared in humans as a natural consequence of inhabitants of the polar regions receiving solar radiation at an oblique angle was true. Should we not then expect the Eskimo people, who live in a place where the sun does not even rise during the long winter months, to have evolved into a race of albinos? Shouldn't we find white-skinned, blonde natives inhabiting Argentina, as well?

Apparently, the shallow angle of the sun's rays only counts if one happens to live in Europe, and particularly in Scandinavia. The skin tone of everyone else on Earth who lived near either the Arctic or Antarctic Circles, rather than becoming pale, remained quite dark. Their hair, rather than becoming blonde, remained black, and their eyes didn't turn blue.

The logical conclusion, when considering this, is that pale skin and blonde hair are not mutations which naturally occur in humans because they happen to inhabit either northern or southern latitudes. They are, according to the best available evidence, an anomaly which occurred in only a single general location. This supplies us with yet another reason to believe that the entire concept regarding the shallower angle of the sun's rays is, basically, nonsense and is disproven by multiple other racial groups which inhabited similar locations.

What we are dealing with here appears to be neither valid science nor a sensible application of the concept of natural evolution. What we are dealing with seems to be scientists who were unable to find an effective way to explain how the Caucasian race came into existence.

An idea about the angle of the sun's rays was the best they could manage to come up with. It was only applicable to Scandinavian-type sun's rays, totally at odds with the fossil record and unsupported by any evidence, but it was all they had.

If they agreed to unanimously get on board with it and limited its discussion in textbooks to a single paragraph, they might be able to sell it to the public. It sort-of-almost seemed like it made sense in a way, if they could prevent people from thinking about it for more than a minute or two at a time…and it was the best explanation they could manage to come up with.

They decided to support it with the full weight of their combined authority: blonde, pale-skinned Scandinavians combined with brown-skinned, black-haired Eskimos and Argentinians. Keep it to one short paragraph in the high school science books and announce that anyone who disagreed with it was obviously not very intelligent and should be ignored by any sensible person.

It was, they proudly proclaimed behind closed doors, a triumph of modern science and a landmark in progressive thought! Best of all, it was backed up by the fact that they would be more than willing to grant anyone who agreed with them and stayed in school long enough official recognition in the form of a Doctorate of Anthropology.

It was, in other words, a perfect solution—except for the part about being utterly ridiculous and completely devoid of any form of supporting evidence, which nobody was likely to notice anyway. And, just as they had guessed, a century-and-a-half has now passed by and it seems that nobody did.

As it turns out, this half-baked, unproven exercise in unscientific desperation is perhaps the least of the problems associated with the universally-accepted explanation of the natural evolution of humanity. Once we dismantle that one, things go from bad to worse as Godzilla's mighty foot continues to descend upon it.

The next thing to consider is the time frame during which all of this is supposed to have occurred. A period of 60-80,000 years is, according to the tenets of the Theory of Evolution itself, far too little for a tribe which wandered north out of Africa to somehow "evolve" into the Swedes, the Chinese, the Eskimos, the Samoans, the Vietnamese, the Apaches, the Mongols, the Indians, the Pygmies and all the other racial groups we see around us in the world today.

First, it must be presumed that there was some compelling reason that such changes in the physical characteristics of humanity would provide them with a survival advantage over others—which there is not. Even had such an advantage existed, the time required for such changes to take place through natural evolution—a slow, torturous process which relies on a series of random, unrelated genetic mutations to occur, should be far longer than 80,000 years. It could be expected to take millions of years at the very least—more likely, tens of millions.

Even so, we are told that it all took place in that short span of time, and that this idea fits in perfectly with the Theory of Evolution— which it doesn't. The simple fact is this: one does not transplant a group of African tribesmen to the area we now call China, wait a few thousand years, then return to find that, for no apparent reason, they have all become Chinese. Not only that, but that all traces that any citizens of Africa ever set foot there had inexplicably vanished into thin air.

I am sure you will agree with me when I point out that such a thing is such a remote possibility that it hardly even bears consideration by serious people. We are not, however, concerned with serious people now—we are speaking about mainstream scientists, who certainly do not appear to be serious about much of anything.

In fact, almost nowhere in the entire world, other than on the continent of Africa, do records exist which speak of the presence of African tribal members who wandered into the area, set up camp, and stayed—eventually to "evolve" into whatever the local variety of humanity happened to be. In some areas, they apparently became Cossacks, in others they became Cambodians, and in still others they became Irishmen. All of this, without so much as a chalk drawing on a cave wall to commemorate the original settlers of the area, and the group which was responsible for establishing the original culture

25

upon which the late-arriving Toltecs, Sicilians and Japanese adopted as their own.

I am all for science, don't get me wrong, but it seems to me that the least it can be expected to do is come somewhere close to making sense and contain at least some indication of cohesive logic. So far, this story we are told about the natural evolution of *Homo sapiens* does neither of those things.

On a side note, it occurs to me that it may very well be useful, at some point, to wonder how exactly so many of these seemingly-oblivious proponents of nonsense ended up being able to earn Doctorates and other post-graduate degrees.

If they are unable even to notice that long hours of darkness and low light have no discernible effect on the racial character of people whose ancestors originated in Africa, it is somewhat surprising they were even able to successfully complete junior high school, much less eventually find themselves to be tenured professors.

Unfortunately for them, an unpleasant situation is about to become much, much worse. Any feeble chances to come out of this chapter with at least some of their pride intact that they may have believed still existed, are about to vanish like the records of Eskimo immigrants to the plains of Nevada did.

Have any among my readers, I wonder, ever heard tell of places known as New Guinea and Australia? I presume that most of you have, unless of course you happen to be associated with the nearsighted, logically-inept mob who call themselves "mainstream scientists".

You may be interested to learn that the shortest distance between Africa and Australia is 6,467 miles, or 10,408 kilometers. By any Earthly estimation, that is an enormous distance—and all of it consists of rough, stormy, treacherous, shark-infested ocean. And—wouldn't you know it—the nice folks who invented the fairy tale about human evolution and migration completely neglected to account for either of those places.

Apparently, they were in a hurry that day. Did they conclude that a hardy band of Africans decided one day that it would be a clever idea to swim out into the shark-infested waters until the shore was well out of sight, in search of lands unknown? Do they imagine that they then somehow managed to make their way across the vast Pacific Ocean, until at last reaching the shores of Australia?

Such an idea would seem to be a product of the same type of logic these people are accustomed to working with. In the real world, however, this idea is nothing but an unusual method of providing the local sharks with a meal, never coming within 6,500 miles of the land of kangaroos and kookaburra.

This leaves only one possible method of transporting a group of settlers from Africa to the distant lands of Australia and New Guinea: they had to make use of either rafts or primitive canoes. Once again, it seems to me that we must ask ourselves whether we honestly believe that we could load enough food and water onto a raft to sustain us for a trip that might take two years or more, and then successfully ride it halfway around the world? Could we do so without a map, and without any reason to think Australia even existed? Could we do it while believing ourselves clever to have left behind an entire continent which was literally spilling over with food?

Can we realistically imagine that they were able to make the journey in primitive canoes, after somehow managing to pack enough food and water into them to sustain four people—two men to do the rowing, along with their wives—for a couple of years or so, and to do this for no apparent reason we can think of? I don't know about anyone else but, speaking for myself, I can tell you this: if I paddled six thousand miles into the ocean without sighting land, I would seriously consider turning around and heading for home. Not that it would ever actually come up, because I would surely have consumed my entire water supply during the first two days of my journey and be long dead by then.

I cannot imagine any circumstances whereby such a journey across the endless waters of the ocean would be considered necessary and undertaken. If it were, I cannot imagine any possibility of it resulting in anything other than disaster, much less the discovery of an unknown continent that lay halfway across the world and that there was no reason to think even existed at all.

The entire concept is beyond foolish and inherently doomed to failure. It simply cannot realistically be thought to have occurred. This means, of course, that if we accept the standard version of human evolution as being accurate, the aborigines of Australia had no viable method of ever getting to those places in prehistoric times.

Even if they had somehow done so, I seriously doubt they would have then proceeded to entirely forget that they had once traveled there from a land far away. The aborigines, however, had no such stories, considering themselves to be the original—and only—inhabitants of Earth.

As for the original inhabitants of nearby New Guinea, the tribespeople who live there are to this day among the most primitive inhabitants of planet Earth. If recent laws had not been passed which forbids it, they would happily still engage in both headhunting and cannibalism. Still locked in the Stone Age, these people would not be able to construct an ocean-going raft capable of making a journey halfway around the world if their lives depended on it. To believe that they had once been able to do so but have for some reason forgotten how they did it and are unable to replicate it even now, thousands of years later, is an idea which is so unlikely it doesn't even bear considering.

There is no ancient culture I am aware of anywhere in the world which has histories that speak of the original inhabitants arriving there on foot after leaving Africa. For the most part, they all seem to believe they have been in that location forever.

I wish I could tell you that we are now finished with our deconstruction/dissection of the standard version of human evolution and migration. Sadly, however, there remains one tiny issue which, for the sake of completeness, must be briefly addressed. It appears in the form of a very famous, universally-accepted line drawing:

Or, sometimes, this:

In some cases, it will look like this:

Or, in still other cases, this:

If they happen to be feeling particularly ambitious that day, it can even be this:

According to this standard depiction of the lineal, evolutionarily-correct progression from apes, through proto-humans, and finally to *Homo sapiens,* there were apparently four major steps involved. Or, maybe, five. Or six. Or seven. Or, going by the illustrations, perhaps eight. By the way, what the hell is that thing on the left supposed to be? A deformed spider monkey, drawn by a four-year old?

They are apparently unable to ascertain which—if any—of those it really is. In fact, these charts contain an error factor of 100% plus-or-minus, in terms of how many intermediate stages occurred between the original ape-form, and modern man.

Imagine, if you will, that you turned in a science test at your local university which contained answers that had an inherent error factor of plus or minus 100%. If the question had been, for example, "2+2=___", your answer could have been either "2" or "8", and still achieved the same relative degree of accuracy which is demonstrated on these depictions of the alleged "progression" from apes to human beings. If the question had been "How many individual States make up the United States of America?" your answer could have been anywhere between 25 and 100, and still maintained an accuracy level which is consistent with that of the scientifically-approved illustrations above.

It goes without saying, of course, that had you entered any of those answers you would have immediately had them marked "incorrect" and been given a failing grade for the test. When it comes to the linear progression which began with apes and eventually resulted in homo sapiens, however, an error factor of 100% is apparently not only perfectly acceptable, it is not even considered to be worthy of making note of.

Though it may by now appear that things could not get any worse for the proponents of the human evolution fairy tale, such is not the case. As it turns out, none of those illustrations has any possibility of being even remotely close to representing the truth. One does not evolve from a chimpanzee-like creature to *Homo sapiens* in four steps. Or five, six, seven or eight. That isn't even enough steps to evolve a proper human hand from any of those alleged precursor species.

What, then, are we to think of those illustrations, which we have all seen hundreds of times during our lives?

Well, it seems to me that the answer to that question is as clear as it is inescapable. Those illustrations, none of which agree with each other and none of which is by any stretch of the imagination accurate, are the anthropologist's version of make-believe—a fairy tale, depicted as reality. Something designed to satisfy the curiosity of small children and the mentally challenged, but without any realistic relationship to reality.

Bluntly, it is a big fat lie. A big fat lie that they told us on purpose—and which has by now become so widely-accepted that it automatically pops into our heads whenever the topic of human evolution is mentioned. All without anybody ever bothering to point out to us that it isn't even remotely connected to the real world.

The problems do not end there. You see, none of the intermediate "steps" which appear on any of the above illustrations represents an actual proto-human species. None resemble a true Neanderthal, Cro-Magnon, etc.

What they are, quite simply, are a series of drawings someone with a pencil and some free time came up with, which purport to demonstrate that each successive step of the progression resulted in an ancestral form which was less like an ape than the previous one, and more like a modern human. Then they inserted them, one by one, in between the drawing of a deformed monkey their kid came up with and a modern human.

Once this had been accomplished, in the time-tested, unanimously-approved tradition of mainstream science, they called it good enough. They then ran off some copies, mailed them out to textbook publishers and our children now had an officially-sanctioned example of complete nonsense, which they would be assured represented established scientific truth.

As noted above, these illustrations have become so prevalent throughout our society that people have come to regard them as *actual scientific evidence*. I have had otherwise-intelligent, educated, full-grown adults produce those illustrations, wave them triumphantly in my face and pronounce that here was all the "proof" that was necessary to nail down the idea of natural human evolution, and end all discussion of the matter.

Really? Some childish pencil drawings, scientific evidence of the ascent of man?

I cannot imagine what the hell these people were thinking of. Do you know what those illustrations provide scientific proof of? One thing, and one thing only: they prove that, at some point, somebody had a pencil and a sheet of paper. That's it. Nothing more.

Try to imagine the reaction I would receive from the scientific community if I were to attempt to utilize a similar pencil-drawing to prove the reality of alien contact. Imagine that I presented the following illustration and asserted that it was based on scientific evidence:

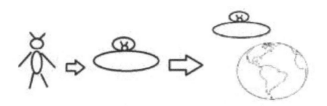

Technically speaking, my depiction is more consistent than that of the evolutionary pencil-sketch artists: it shows one—and only one—alien spaceman present during each stage of the putative interstellar voyage. Even so, I would rightly be laughed out of town, were I to attempt to utilize this as some form of evidence or proof. With that in mind, should we not then be laughing those responsible for creating and officially endorsing the illustrations depicting the progression from ape to humans out of town as well?

It gets even worse, as far as their official illustration is concerned. You see, there is no evidence whatsoever that any of these so-called "human ancestors" are related to each other at all, much less directly related to homo sapiens. There are no intermediate forms in the fossil record showing that any of them naturally evolved into any of the others, and none showing any of them evolving into modern humans.

The evolutionists needed a clear progression, a timeline of some kind…but, after studying the fossil record, none was there to be found. What they decided to do was line up the primitive hominids which were part of the fossil record in order, according to when it is believed they existed. This gave the impression that each species had naturally evolved into the next on the list over time, and that this had been demonstrated by solid scientific evidence.

This was in no way true. In fact, assembling them as they did on an alleged "ascent of man" chart was unscientific, unsubstantiated by the fossil record, intentionally deceptive and outright dishonest.

What occurred is that time after time, throughout the past several million years, these primitive hominids suddenly appeared in the fossil record, fully formed, and the species which had come before them vanished. One day there was no such thing as a Neanderthal…the next, there they were, completely "evolved" into an entirely new species, making stone tools and living in small social groups. There was no sign at all, at the time, of the species we now refer to as Cro-Magnon.

Then one day, suddenly and without any warning, Cro-Magnon arrived on the scene, fully formed, using primitive tools and living together in small bands or tribes. There is no indication in the fossil record that Neanderthal somehow evolved into Cro-Magnon, and no legitimate scientific reason to believe that this was the case.

Time after time, as best we can determine from the fossil record, this cycle was repeated as various primitive, ape-like creatures suddenly came into existence. As they did, the species which had previously been thought to represent the most highly-evolved example of what were assumed to be the descendants of apes conveniently managed to completely disappear. They became extinct, and the most current version of hominid-kind took over in their place.

There is also no evidence, by the way, which demonstrates that any of these primitive forms evolved into modern man. Not one of our supposed "ancestors", including the so-called "nearest relative" to *Homo sapiens*, had so much as a single human bone in its body.

That is not what I would consider to be a "close relative" at all. In fact, I would classify it as something which may in fact be entirely unrelated to modern *Homo sapiens*. At the very least, if it is in fact a legitimate ancestor of ours, there must have been at least hundreds—and, more likely, thousands—of intermediate forms.

The transformation of the creature we are told represents our closest direct ancestor to homo sapiens is far more than a matter of a few simple evolutionary changes. Instead, it requires a complete, wholesale anatomical restructuring of the body.

This cannot be accomplished in one, or even several, generations. It requires an enormous number of random, unrelated genetic mutations to occur, and the passage of a correspondingly enormous amount of time. There must have been numerous pre-human forms which existed between this so-called "closest ancestor" and us, which were slowly evolving the skeletal structure and other characteristics of what would someday become *Homo sapiens*. As is the case with beings such as Neanderthal and Cro-Magnon, we should reasonably expect to find them represented within the fossil record.

Evidence that such intermediate-stage beings ever existed anywhere in the world is, however, entirely absent. The chances of this happening—thousands of intermediate forms leaving no trace of their existence in the fossil record, while only completely-developed species are found—are so remote as to be inconsequential. The odds that this would happen repeatedly, occurring with every new iteration of a primitive, ape-like species, approach the vanishing point.

This is the reason we occasionally hear people speak of the search for what is described as "the missing link". It refers to a species of hominid which can be conclusively demonstrated to represent a form which was, at one point, this closest ancestor but was slowly evolving into what would someday become modern man.

No such "missing link" has ever been located. I do not believe it ever will be located, because I do not think it exists to be found. It should be pointed out, as well, that far more than a single "missing link" is necessary, if we are to make the orderly progression from what we are told is our closest ancestor to ourselves. As was noted above, we should naturally expect to find evidence of hundreds, if not thousands, of intermediate forms which existed as they slowly underwent the radical, profound evolutionary transformation which was required to produce modern man.

Scientifically speaking, there is no compelling reason to assume that all these many intermediate species existed at one time, but for some unknown reason are completely absent from the fossil record. Such an assumption defies common sense. It also violates the most basic principle of science, which states that scientific conclusions must be based upon solid, verifiable evidence of some kind. What we have, instead, are a series of conclusions which are based upon the idea of creatures which is it believed must have existed at some point, but which there is no compelling evidence to demonstrate that any of them did.

This is a profound—and, seemingly, fatal—problem which can be found throughout almost the entire fossil record of Earth. It is not limited to just the evolution of modern man, it also applies to the evolution of most other species of animals in the world.

That problem does not fall within the parameters of this manuscript and will not be discussed here. Besides, mainstream science has already got that situation covered. If one is willing to accept something as being true, when 99.999% of the fossil evidence which *must* be present appears for some reason to not exist, and the idea that a single-celled amoeba could someday evolve into a blue whale, an ostrich, a camel and William Shakespeare, everything falls right into place.

This lack of evidence is the reason this entire school of thought is called the "theory" of evolution, rather than just "evolution". If the evidence that something occurred is notably absent, there is no sound scientific reasoning which allows us to make the presumption that it did, in fact, occur. To make such a presumption, regardless of the lack of any evidence to support it, is a violation of the basic principles of scientific methodology. It becomes, at that point, nothing more than a matter of guesswork and conjecture based upon evidence which, to the best of our knowledge, does not exist.

There is no other field of science in which this type of unsupported reasoning would be found acceptable, much less found sufficient to base legitimate scientific conclusions on. Without the presence of evidence and proof, science itself cannot even exist.

There is also the matter of gigantic skeletons, all apparently human, which have been unearthed at various times throughout history. As we might expect, there have been instances where skeletons of giants were reported to be found, which were later exposed as hoaxes—and we are not interested in hoaxes. But it is also true that many giant skeletons have been reliably reported and there is no reason to believe they were not legitimate finds.

Genesis 6:4 states that "There were giants on earth in those days; and also, after that, when the sons of God came in unto the daughters of men, and they bear children to them, the same became mighty men which were of old, men of renown." While I am not someone who considers everything in the Bible to be literally true, there is no question that many references in the Bible are known to be historically accurate. The presence of giants seems to be something there would be little reason to invent, since there appears to be nothing for anyone to gain by doing so. It is also corroborated by several ancient skeletons which are known to exist, so I do not see any reason to presume the biblical statement is untrue, and it stands to reason that it could very well be accurate.

Please note that the word "giants", as referred to here, does not include cases of gigantism or acromegaly, both of which result from physical disorders in which the body receives excess growth hormone. This results in individuals such as Robert Wadlow, who reached a height of 8'11", "Andre the Giant" (7'4") and others who suffered from one of these conditions. These individuals were not examples of "a race of giants", and as such are not relevant to this discussion.

Those cases aside, there are cases which appear to represent legitimate humans who were of gigantic stature. In 1912, for instance, 200 burial mounds were located at a dig site near Lake Delavan, Wisconsin which was overseen by Beloit Collect. The mounds were found to contain apparently human remains which ranged from 7'6": to 10' tall. This find appears to be legitimate and is only one of dozens of finds which have been reported in the American Midwest between 1850 and the present day. They are by no means the only reports of giant skeletons which exist, and which appear to be accurate depictions of legitimate skeletons.

A 19-foot skeleton was found in 1577 A.D. under an overturned oak tree in the Canton of Lucerne. A 23-foot tall skeleton was reported found in 1456 beside a river in Valence, France. In the late 1950's, a 15-foot tall skeleton was found in Turkey. Human footprints 22 inches in length were found in 1932, in the gypsum near White Sands, New Mexico. In 1879, a 9'8" skeleton was found in a burial mound near Brewersville, Indiana (Indianapolis News, Nov 10, 1975).

In 2002, National Geographic reported that a dozen "cyclops" skeletons had been recovered in Greece, which stood between 10-and-a-half and 15' tall.

This is only a small sampling of the examples of giant skeletons and footprints which have been reported. There are dozens more, from all over the world. The largest reported human-appearing skeletons were found by the Carthaginians between 650 B.C. and 640 A.D. Two skeletons were found which were 36 feet in height, and an earthquake later uncovered a third. By comparison, a six-foot-tall man would not even come up to the knees of a being this size.

You will not find these skeletons in any museum, however. True to a long-standing tradition in conventional science, evidence which conflicts with currently-held beliefs, including giant human skeletons, often conveniently disappears soon after being turned over to museums for safekeeping.

This very much applies even to the Smithsonian Institution, a place which is often called "the black hole of evidence", where many such skeletons have been given into their custody and never seen again. It has even been reported by a former employee that the Smithsonian at one point loaded up a full barge of such giant skeletal remains, sailed it out into the Atlantic Ocean, and dumped all of them over the side into the sea!

There is no question that science has often been far more concerned with protecting its currently-fashionable ideas and beliefs by eliminating the evidence, than it has with examining the evidence in an unbiased manner in a search for truth. It has clearly been the case when it comes to the subject of human evolution, among many other things.

Why would we choose to invest our beliefs in ideas which have motivated scientists to destroy evidence which contradicts them, rather than be forced to admit they were mistaken about it in the first place? I can think of no sound reason to do such a thing. Considering that, along with the issues identified earlier in this chapter, I do not see any reason to believe that their story about the natural evolution of man is worth much more than the ink it took to print it.

It must also be pointed out that skeletons of this size clearly do not represent examples of *Homo sapiens* at all. They must represent either distinct species, none of which are explained—or even acknowledged as ever having existed—by conventional science...or they must represent the fossilized remains of alien beings.

We must also consider the fact that the conventional theory of human evolution has no means of accounting for, or explaining, the fact that multiple alien races are known to exist which are either as human as we are or are extremely close to us in appearance and function. If humanity evolved here on Earth completely independently of those alien races, the chances that several other races would evolve in such a manner as to be all but identical to us is vanishingly small.

There must be some way of explaining this situation. For that matter, there must be some approach which is able to account for *all* the problematic issues which arise with the conventional theory, and explain them in a logical, compelling manner. Clearly, the theory we are given fails to do any of those things. It must, therefore, be abandoned. This is something which should have been done long ago, in my opinion, but never has been.

CHAPTER 4: HUMAN ORIGINS

*"When I reach for the edge of the universe, I do so knowing that
there are times when, at least for now, one must be content to love
the questions themselves."*
-- Neil deGrasse Tyson

*"Boy, you're gonna carry that weight
Carry that weight a long time."*
--The Beatles

Having illustrated the multiple seemingly-intractable issues associated with the theory of human evolution, I hope I have demonstrated that there is no compelling reason to think it to be anywhere close to accurate, or to continue to invest our belief in it. I thank the readers for their kindness and patience in allowing me to include my thoughts as I have. We can now proceed to examine the concepts I referred to a moment ago.

It turns out that natural evolution and Creationism are not the only two possibilities, when it comes to the origin of man. There are additional possibilities to consider. All of them, however, are automatically rejected without discussion by conventional science. That does not imply that they are unworthy of consideration—rather, it simply exposes another blind spot which burdens (and slows the progress of) mainstream science.

The most surprising—or, perhaps, ironic--aspect of the refusal of the scientific establishment to consider these ideas is that they appear to be able to tie up at least some of the loose ends left behind by the standard theory. They can also provide us with viable solutions and answers to the problematic issues mentioned previously.

The first of these additional possibilities deals with the idea of long-term genetic engineering, carried out by alien scientists, right here on Earth. The only thing necessary for one to be able to seriously consider this idea is the acceptance of the possibility that advanced alien beings have been present on and around this world since prehistoric times.

It is a fact that there has never been a time in the recorded history of humankind, including cave drawings which have been dated to at least 12,000 years old, that the presence of alien beings has not been recorded by our ancestors. There is no good reason, therefore, to think that aliens were not here prior to that, and it seems only sensible to presume that they almost certainly were. There is no reason to believe that they could not easily have been present here *long* before the earliest cave paintings.

It does little good for debunkers to claim that those big-headed, almond-eyed humanoids drawn thousands of years ago are intended to represent grasshoppers—we are not stupid, and we recognize drawings of what are clearly alien greys when we see them. When we see ancient drawings that look the same as modern-day alien spacecraft, we recognize them—saying they are just supposed to be "the sun god" is an insult to our intelligence and something we know better than to believe.

If advanced alien beings were here as far back into the past as several million years ago, we can easily construct a scenario by which we can produce an entirely possible, logically consistent explanation for the origin of humanity. We can even construct one which is able to fill in the gaps and answer some of the questions that conventional evolutionary theory is incapable of adequately addressing.

First, we know that several alien races, including the Pleiadeans, Nordics and others, bear a striking physical resemblance to humans. The chances of this occurring by coincidence are surely miniscule, if all these races—including ourselves—are the product of entirely independent evolutionary progressions.

It may be that the general humanoid form (two arms, two legs, one head, two eyes) is so efficient that it evolved naturally on many different worlds. That is a very different thing, however, from discovering multiple alien races whose evolutionary development so closely mirrored our own that it is often difficult to tell the difference between them and us.

It makes much more logical sense to seriously consider the idea that we are, in fact, close genetic relatives. Their DNA could have been combined with that of already-present prehistoric apes at some point and used to produce what we think of as modern human beings. In other words, *they* don't look like *us*, *we* look like *them*—because we carry their DNA within us.

Interestingly, this could be seen to have a rather striking correlation to the verse in Genesis which describes the creation of man. In that account, the Lord said, "Let us make man in our own image". If we allow for a mistaken perception in ancient history, it is sensible to consider the idea that advanced alien beings could easily have been confused with God in those days.

If so, we could understand how this verse—which otherwise cannot be reconciled with science—could turn out to be close to the truth. Alien genetic engineers, creating a hybrid being by crossing their own genetic material with that of apes which were native to Earth, could be said to have literally created man in their own image.

If it happened to be an alien race which did not resemble us that was responsible for producing such a hybrid form, it is entirely reasonable to think it likely that they also had access to the DNA of alien races which *did* resemble us and used it. Either of these basic scenarios would provide a theoretically possible explanation for the physical resemblance between *Homo sapiens* and several alien races which are known to exist.

Genetic engineering is undoubtedly a complicated project under the best of conditions. If one wishes to combine their own genetic material with that of a creature from another world, it must surely increase the difficulty of success by orders of magnitude. We would not expect a first-generation product of genetic engineering which was carried out on primitive ape-like creatures to result in anything even close to modern humans—and, from the looks of things, it seems that it did not.

Many people, relying on the interpretations of ancient Sumerian symbols produced by Zechariah Sitchen, believe a race known as the Annunaki was responsible for the creation of humanity. I have no way of knowing whether that is correct. It may not be necessary to our present discussion to know precisely which alien race or races may have been responsible for our creation, if such a thing is indeed the case.

41

For the purposes of this theory, it would not appear to make much significant difference which alien races may have been involved, so long as they ultimately were able to get ahold of DNA from one or more races which bear a strong physical resemblance to ourselves. That being the case, let us postpone the matter of attempting to identify them specifically, and think in general terms for the moment.

Please bear in mind that much of what I say in this chapter is speculative and unproven. Even so, it seems unlikely to me that, even in the worst case, it will end up looking nearly as battered and bruised as the conventional theory of human evolution that we examined in the previous chapter.

It is my understanding that the planet Earth is located on an ancient trade route which is traveled by numerous alien races and has long been used as a stopping-point by many of them—a place where they could rest, refuel and, if desired, collect minerals or biological specimens. A trade route such as this could quite easily be something which has been in constant use for millions, or tens of millions, of years.

There is no reason I am aware of to presume otherwise, since it certainly appears that many of the alien civilizations which either do or have visited our world are quite ancient. Rather than utilizing trade routes which have been in place for hundreds or thousands of years, as is the case here on Earth, they could well have established trading partners and routes which extended across the galaxy and lasted for millions of years.

If it is true that Earth is located on or near a well-established trade route, it would also negate an idea which is often voiced by skeptics. The thrust of this idea is that there is little chance our world would have been noticed, considering the huge number of stars in our galaxy.

It is safe, in my opinion, to believe that races whose civilizations can be traced back for literally millions of years are quite unlikely to use the same time frames we are in the habit of using when making their long-term plans. We often come up with plans which are intended to produce results within days, weeks or years. The longest human plans may be designed to cover a period of a few decades.

That would be unlikely to be the case with these alien civilizations. While we think ahead in terms of, for instance, a five-year plan, they may well be operating according to plans which cover a half-million years. For all we really know, their long-term plans may potentially require millions of years to achieve their intended purpose.

If one intends to start with primitive, tree-dwelling Terran proto-apes, and eventually end up with a species such as *Homo sapiens*, it is clearly a project which is going to require a significant amount of time before it comes to fruition. If the alien race in question only stopped here periodically while engaged in interstellar trading voyages, rather than building permanent colonies here, it seems possible that such a project may well require not hundreds or thousands of years, but *millions* of years.

Again, this is what appears to have been the case. Please keep in mind that I am not attempting here to find a way to make the theory correspond with the fossil evidence. I am, rather, offering a speculative idea which can in fact be made to fit nicely with the fossil record, without having to invoke ideas which either do not make sense or are not realistically possible enough to take seriously. This one is worthy of taking seriously, because—as I mentioned at the beginning of the chapter—it can account for both the gaps in the fossil record, the racial diversity within humanity, and the time factor which is so problematic to the standard story of human evolution.

What if alien genetic engineers "upgraded" an already-present primate during one of their regular interstellar journeys, then left it alone for a few thousand years or so, to see how well it was able to survive here in its new form? Then, after returning much later and seeing it still here and apparently doing well, they went ahead and modified it again, making it somewhat more intelligent and capable than the previous version?

What if they repeated this process again and again, over a period of hundreds of thousands of years? Their eventual goal may have been to create a being which was intelligent enough to survive on its own—and, perhaps, to mine some gold for them (a la Sitchen's theory) but not smart enough to ever pose a realistic threat to them.

When they returned to find that their latest "upgrade" was successful and thriving, they may have eliminated the previous model, or perhaps the newer model did that on its own. If something like this scenario occurred, it would explain the absence of intermediate forms in the human ancestral tree, as well as the significant differences between the primitive proto-humans we find in the fossil record.

Please note that it is not necessary for the scenario I just described to be accurate in all its details to account for the difference in form and the lack of a "missing link". All that's needed is for the general sequence to have occurred in some manner, for a reason which is perhaps known only to the aliens themselves. There could be any number of reasons—including pure experimentation—it was decided to undertake a long-term program of genetic engineering here on Earth.

If the alien race or races in question did indeed think in terms of epic time scales, which could easily be thought to be the case when dealing with ancient, long-lived civilizations, there is no reason they'd have needed to hurry. If the result was acceptable to them, it may have made no real difference to them in practical terms whether it took ten thousand years or a half-million years to produce the finished product to their satisfaction.

If that finished product required multiple "tweaks" along the way, it may not have made much significant difference to them whether the number of modifications proved to be ten or eighty. They were still going to be traveling the same route between the stars either way, and it would take however long it would take. If they happened to have a civilization which was, say, thirty million years old, whether their project required 20,000 years or 200,000 may not have mattered to them overly much.

It is also possible that their original intention was to create an intelligent race which would provide them with a source of young, strong, highly-diverse DNA from which to produce hybrids of their own race. They may well have understood that such hybridization would someday be necessary for their own survival.

If such was the case, they may have never intended for their prospective hybrid-match to be fully completed until they anticipated having a need for it. That would have ensured that our DNA was as young and strong as possible when the time came to use it to produce hybrids—an ideal situation, as far as they were concerned.

It is also possible that they may have simply collected some primitive ape-like creatures here, transported them back to their home world and proceeded to work on engineering them there, at their leisure. When a new species they felt had potential was produced, they could have moved them back here and dropped them off, waiting to see how they fared even as they were working on the *next* upgrade.

There is another idea which can easily be brought into such a theoretical scenario, as well. After at last arriving at Homo sapiens, they may have decided to create multiple versions of their "finished product", each of which was slightly different to the rest, but all of which remained true to the general model.

In other words, comparatively minor modifications to the DNA of *Homo sapiens* is all that would have been necessary to produce all the different human races which are in evidence today. They could have separated these races geographically, making it difficult for them to easily interbreed until they acquired sufficiently-advanced methods of transportation, far in the future, and waited to see how each version developed.

An additional—and, it seems to me, quite compelling—point in favor of this idea is that neither Creationism nor evolutionary theory has any possibility of explaining why humanity bears such a close resemblance to some alien races. Genetic engineering, however, can do so.

This is all speculation, of course—I have no way of knowing what their motivations may have been. But, regardless of which variation of this project may have taken place—including those I have failed to consider—the result would still account for the otherwise impossible-to-explain gaps in the fossil record.

It would explain the lack of intermediate forms. It also gives the possibility of providing an ancestral lineage which would be both logical and practical, from the point of view of an alien civilization which traveled through our solar system regularly as a matter of habit or on trading missions.

All that is necessary for these or a similar scenario to have played out is for intelligent alien life to exist, to be able to get here, and to possesses genetic engineering technology. These things are already known to be true regarding at least some alien races we are familiar with.

In terms of providing us with a much less problematic solution for the origin of humanity, it doesn't much matter what the motivation of the aliens may have been. Regardless of which scenario may have been in play, we are still left with a better explanation than we had before—not to mention one less inaccurate, deceptive and intellectually insulting pencil diagram of monkeys inexplicably morphing into modern man.

It also eliminates the need to concoct an explanation of how and why blacks "evolved" into the rest, and why the fossil record does not reflect that this ever happened. No need to invoke slanted solar rays leading to pale skin, and no need to explain how being Chinese or Norwegian resulted in such a survival advantage that all traces of black inhabitants in those lands vanished into thin air.

The key issues which the conventional theory of human evolution cannot resolve would all be eliminated, simply by being willing to believe in an alternative possibility which we already know to be entirely possible and logically feasible. Otherwise, we end up being stuck with a theory of evolution which is directly contradicted by the absence of fossil evidence to support it, and disturbing logical issues which cannot easily be overlooked or dismissed.

Standard "Creationism" cannot be reconciled with science. Conventional evolution cannot be reconciled with the evidence. This makes them both incompatible with legitimate scientific methodology, and therefore unacceptable. Alien genetic engineering alone solves the problems they cannot solve and answers the questions they cannot be properly made to answer.

Of the three possibilities, only one appears to be compatible with the fossil record. Only one can provide a technically feasible, logically-consistent explanation for the uncanny resemblance of humanity to certain alien races.

When all is said and done, the idea of alien genetic engineering being responsible for the origin of humanity provides us with the only potential explanation which can be described as fundamentally sound and scientifically credible. If there are three possibilities, and two of them cannot be reconciled with proper science, the third must represent the truth.

On a heavily ironic note, if this idea of the origin of man is correct, it would also mean that, when all is said and done, the Creationists were closer to the truth than were the evolutionists. Science would never forgive me for making such a statement. This is how I know it will never accept my ideas as valid.

That does not mean they are incorrect—it just means science is unable to come to terms with them. Instead, it will inevitably choose once again to hide its pointy little head in the sand, hoping that doing so will make me go away and leave them alone.

That is the best case I can make for the genetic engineering explanation of human origins. Still, it doesn't satisfy me. It fails to account for, or explain, the fact that multiple alien races are known to exist which are either as human as we are or are extremely close to us in appearance and function. We must therefore consider additional possibilities, hoping to find at least one which can provide us with an explanation for those problematic issues.

Over the past several years, it has been my good fortune to have become a close friend of a man called Jerry Wills (www.discoverytv.com), who resides in Phoenix. Jerry is, by all accounts, a native of the Tau Ceti star system who was transported here as a young child and presented to representatives of our military black operations forces. He was subsequently turned over to a childless military couple and raised as their son.

I have no way of proving that this is what occurred, of course, but I know Jerry quite well and neither I nor anyone else I am aware of has been able to find any reason to doubt this claim. I accept it as true until and unless such a reason is someday found (which I seriously doubt will ever happen).

Jerry is as human as you and I are, in all ways which matter. He is taller than most of us who were born on Earth are, standing six feet nine inches. He is also more intelligent that most of us here can claim to be, possessing what is surely genius-level intellectual capacity and being highly knowledgeable in multiple sciences and other fields of study. In the next volume, I will include the transcripts of some conversations I had with Jerry, which I think the reader will find to be quite interesting and informative.

Jerry also possesses the ability to heal others, using only the power of his mind. That statement is not a guess or an unsupported claim of any kind—in fact, he has demonstrated his ability on numerous occasions. He has rapidly healed people of injuries which were beyond the ability of modern medical science to effectively treat. This includes bringing a man with a severe head injury out of a coma on one occasion, at least two cases of healing blindness and healing third-degree burns within only minutes.

I can attest to the fact that Jerry Wills, using nothing but the power of his mind, completely cured me of both sleep apnea and habitual snoring, as well as doing much to rectify my symptoms of congestive heart failure. He accomplished all this within the space of a half-hours' time, while I was visiting him at his home. My regular medical doctor was unable to do any of these things for me, no matter how he tried.

I regularly come across individuals who claim to be of alien birth, as I will discuss in a later chapter. I have so far not found any of their claims to be compelling, and I do not accept them as being true. It is possible that some of them are true, of course, but as of now I have seen no evidence which would convince me of it.

Jerry Wills is the sole exception, at this point. I know him well, he is a close friend of mine, I have been to his home and we've hung out quite a bit. In all that time, I have found no reason to distrust him and to the best of my knowledge he has always spoken the truth to me.

I accept his claim of originating in Tau Ceti as true. It seems that he was intended to be a gift to this world, a healer who would spend his entire life here, living as we do, part of human society.

The reason I mentioned Jerry Wills is that it has long been known that alien visitors who appear to be humans who have adapted over time to other worlds and environments have been living among us here on Earth. In addition, there are multiple instances of what certainly appear to be highly-advanced human civilizations here which existed many thousands of years ago.

The most logical explanation for the situation is that humans have traveled here voluntarily and established societies on Earth long ago. Rather than being a species which originated and naturally evolved here, it certainly appears that modern homo sapiens are instead a race which has been imported to this world from some far-distant location, long ago.

In fact, it would appear likely that humans from other worlds have come to Earth and colonized certain areas of it on at least several separate occasions. Some of those alien colonies or civilizations appear to have been destroyed in conflicts which arose between themselves and other aliens which were hostile to them.

When one considers tales of cities such as Atlantis, highly-advanced civilizations such as the Toltecs, the Mayas and others, this is an idea which makes logical sense and appears likely to be corroborated historically as well. Empires such as the Toltec and Maya, however, were totally forgotten by the native residents of the areas where they once existed. All they knew about them was that they had occasionally located large cities which once belonged to them but had long since been overgrown by the jungle. Any memories of them among the local tribes that now occupied the same area apparently vanishing completely.

This is something which does not make logical sense. The occupants of those lands were often primitive stone-age tribes, although sometimes they were members of civilizations which sprang up thousands of years later, such as the Aztec Empire. To have an entire group of people who were dispersed over a vast area of land completely forget that these ancient, highly-advanced cultures had even existed is difficult to understand.

It would be somewhat like having New York City destroyed in a war, then finding that nobody in the area was even aware that it had ever existed. I cannot imagine such a thing happening in the real world…unless those ancient civilizations, along with their entire populations, were either wiped out wholesale or decided to move back to their original planet of origin. Perhaps it was a combination of both those things, depending on the location in question and precisely who had initially established themselves there.

Whatever the case may have been, I think we can safely presume that empires such as these would not have been utterly forgotten by all those who lived in the area after them…unless they had never been aware of them in the first place. The only way I can think of for such a situation to come about would be if the entire populations of those ancient empires suddenly vanished for some reason. If that occurred, the ruins may have eventually been covered by jungle and, at some point, inhabited by people who wandered in later from somewhere far away.

Did humans travel here from some far-away planet at some point in the distant past? If so, it would help to explain why we seem to so closely resemble some of our extra-terrestrial visitors. This aspect alone would seem to make the idea more likely to be accurate than either standard Creationism or standard Evolutionism can claim to be.

It is not an idea, however, which is able to provide answers to the complete list of questions which must, in my opinion, be addressed during an honest inquiry. At least a couple of issues occur to me which it does not seem able to sufficiently address.

One of these issues deals with the question of why, if advanced races of beings which were closely-similar to us were responsible for our presence of Earth, primitive Stone Age societies existed here at all. If our direct ancestors were able to travel here from some distant solar system, there would be no obvious reason for such a situation to have occurred.

It is very clear from the fossil record, however, that primitive human societies which utilized tools made from chipped stones and carved wood existed wherever ancient human fossils have been unearthed. Why would this be true?

The second issue which comes to mind has to do with the fact that, during the first six weeks of its existence, the human fetus demonstrates a physical form which is distinctly reptilian in appearance. During this time, its form is not in any way like that of the human it will eventually become.

What is the reason for this? Though it has long been a known and accepted fact throughout the scientific world, an explanation for this apparent anomaly which is, in my view, sufficiently logical and compelling has never been put forth by the scientific community. Such an explanation *must* exist *somewhere*. The question then becomes "where is it, and why hasn't it been found?"

Regarding the first issue, it does seem that there are several ancient civilizations that have been unearthed which were highly advanced. They appear, in fact, to have been much more technologically advanced than can be reasonably explained by ascribing them to be the work of the ordinary humans who existed at the time.

The ruins of some of these civilizations have been found buried in the overgrowth of dense jungles on both the South American and African continents. Others have been located deep beneath the surface of the oceans, on the sea floor. All of them appear to have been far more advanced than any comparable human civilizations of their time.

We have no way of demonstrating conclusively that any of these civilizations were the result of human-like (or human) alien races which at various times chose to inhabit Earth but later—whether voluntarily or not—moved on to other locations. It is, however, certainly possible, and I think it is well worth considering.

If this were the case, it would go far toward explaining a physical resemblance between them and us which it seems is far too great to be the result of entirely different evolutionary cycles on worlds which were separated by vast cosmic distances. Since this is something which none of the conventional stories about human origins can address in any way at all, much less do so in a manner which is both logical and compelling, that fact alone makes this idea worthy of careful consideration by researchers.

It does not, however, explain the presence of primitive human civilizations in ancient times. It should be remembered, as well, that human civilizations which literally utilized Stone Age technologies existed throughout all of what is now the United States and Canada when European settlers arrived here a scant few hundred years ago.

In addition, primitive Stone Age tribes are known to exist in some areas even today, deep in the jungles of South America, Africa and New Guinea. None of these tribes show any indication of being the descendants of some highly-advanced alien race which was able to travel through interstellar space and construct bases here on Earth many thousands of years ago. This is also true of the Native Americans who were present across the breadth of the United States and Canada only a short time ago.

How, then, can this incredibly improbable resemblance in form be explained? Are Homo sapiens as we know them the result of a breeding program carried out by another extra-terrestrial race, which utilized the DNA of human-like visitors and then turned the products loose on Earth to fend for themselves? Or perhaps even turned them loose and then hunted them for food, in ancient times (as it is believed some still do to this very day)?

Those are questions I do not have the answers to, nor am I aware of any method available to me which can determine whether some or all "human-like" E.T. races spend their first several weeks as a fetus in reptilian form. Perhaps someone who works deep underground, at a highly-restricted military facility, possesses the answers...but I do not.

I asked my friend Jerry Wills, whom I believe to be the only legitimate person of extra-terrestrial origin I know, if he could answer the question about the reptilian form of newly-conceived fetuses among the residents of Tau Ceti. Did they, too, show this odd characteristic, or did they have a human appearance the whole way through?

Jerry replied that he simply didn't have this information—it had never been given to him—and remarked that just because his origin was from a faraway place, does not imply that he would have information such as this at his disposal. That seems to me to be an entirely reasonable answer, and I take him at his word on this. If he knew the answer, I'm sure he would have been happy to share it with me. If he doesn't know it, he just doesn't and that's the way it is.

If the answer had turned out to be no, that they do not take on a reptilian appearance during the first six weeks of gestation, then we could perhaps theorize that Homo sapiens are the result of genetic engineering on the part of another race. Perhaps it was an alien race such as the Draco, who utilized some form of their own DNA and then caused the human DNA to be activated after six weeks in the womb, and produce what we consider to be modern humans, who were then released into the world for reasons of their own.

Lacking such knowledge about the reproductive details of extra-terrestrial races, however, does not allow us to presume that this was the case with a high degree of confidence—it would be nothing more than a guess on our part. That may be the most we can come up with, at this point.

I am not, however, paid to guess about such things. Furthermore, I *hate* having to guess about important matters, and risk getting it wrong and adding confusion to the mix. I feel it is better to simply admit that there are many questions I cannot provide the answers to and leave it at that.

Like all of us, I would certainly love to know the ultimate truth regarding human origins. It turns out, however, that this is one of those secrets which is beyond my ability to decipher, at least for now.

Even so, I hope to have provided the reader with some possibilities which are at least more likely to lead to the answers we seek than those which are espoused by conventional science. I also hope I have demonstrated why the "answers" conventional science supports appear to be fatally flawed and should probably be either completely revised or discarded altogether.

As I have made progress in my research by standing on the shoulders of those who came before me, I gladly offer my shoulders to researchers of the future to make use of, in their own time and way. This is how progress is made, step by painful step, and at some point the answers we seek will hopefully be made clear.

When we find that the answers we seek are beyond our grasp, loving the questions themselves can be enough to provide us with the inspiration to continue our journey. We cannot possibly wish to have had the opportunity to live in more interesting times than these, or to have participated in a more fascinating journey than the one we are on. This is a priceless blessing, if only we take the time to step back and realize it, for our journeys will someday become our lives. Seek the answers, my friends…but, when they cannot be found, be content to love the questions themselves.

Note to the University of Washington Anthropology Department: Just go ahead and mail me the doctorate. There is no need for an elaborate ceremony, and the idea of having all of you kneel before me and proclaim your collective submission is not something which my ego requires. Even if it were, I would never want to embarrass you like that in front of your little friends.

I recognize that the contents of the previous two chapters—my doctoral dissertation—is somewhat shorter than you are accustomed to. It is unfortunate the other applicants were unable to find a way to make their case in less than forty pages, as I have done, but that is just the way things turn out sometimes. It is an anomaly which is best taken up at another time, if at all.

I assure you that I will not be seeking to replace you at your jobs or anything, so there is no need to worry about that. In fact, you couldn't pay me enough money to spend my days comparing fossilized bones with each other and then drawing the wrong conclusions about them afterward, as you apparently do.

*I understand that some people can make a career of doing this, but I would consider it to be a waste of my valuable time. I prefer to simply spend a few quick minutes correcting and grading your essays and then move on to something more interesting, like...well, pretty much anything, I suppose. ***

To be honest, anthropology is not even something I am particularly interested in. It just happened to come up while writing this manuscript, so I thought "Well, okay, why don't I just go ahead and nail it then."

At any rate, just mail me the doctorate, and we'll call it good— everybody wins. Thank you in advance for your prompt attention to this matter.

*Bad news: there will be no valedictorian this year. In fact, the entire graduation ceremony has been cancelled and I have recommended that you be asked to reimburse the University for the total amount of your annual salaries. I realize that you will be upset to learn of this, but you would expect no less and I am confident that, were our positions reversed, you would do the same thing for me.

CHAPTER 5: OF DEMONS AND ANGELS

"For we wrestle not against flesh and blood, but against principalities, against powers, against the rulers of the darkness of this world, against spiritual wickedness in high places."
--Ephesians 6:12

"I'm praying to the aliens!"
--Gary Numan

Many people believe that the beings commonly described as aliens or extra-terrestrials are literally angelic or demonic entities. It is understandable why this is so, and understandable why such an assumption might be made. This is a complicated aspect of the alien contact phenomenon to discuss under the best of circumstances. It is made even more so by the wide variety of religious belief systems which are present on our world, and our individual interpretation of them. It seems safe to observe that one size most emphatically does *not* fit all when it comes to the matter of religion.

It also seems clear, both from history and from looking at the modern world, that a single religious doctrine or text can be interpreted in radically difference ways by different people. The same words can be—and often have been—construed to mean wildly different things, depending upon the motivations and worldview of the individual. The New Testament, for example, has been interpreted to teach that the taking of human life is a sin. It has also been utilized as an excuse to launch wars of aggression and to sentence people to death without the slightest trace of guilt or remorse for doing so.

If we are to gain a true and proper understanding of the non-human entities which are clearly present on and around our world, it is necessary to consider religion in general and the alien contact phenomenon separately. By doing so, we can vastly aid our understanding of the overall situation, while at the same time integrating the subject of religion in a careful and methodical manner when it becomes necessary to do so.

Without this initial separation, however, the situation quickly becomes confused. Our perceptions of these extra-terrestrial visitors rapidly blurs, morphing into something which cannot be made to agree with the information in the database.

I do not mean to imply that religion and alien contact are unrelated, nor do I believe that to be the case. I mean only that the two subjects should and must be considered carefully from a detached perspective and not mixed together without first understanding the reasons that this may be necessary.

When considering whether the aliens we are dealing with are angelic or demonic entities, the matter can be settled in a straightforward manner simply by examining the matter of three-dimensional reality. This, it seems to me, is the most useful and helpful thing to do first.

It is a fact that the bodies of dead aliens have been recovered on numerous occasions from the debris of crashed alien spacecraft. These bodies have been found to represent a variety of alien races and types. They have been carefully examined by highly-trained medical doctors as well as other professionals and have been subjected to autopsy procedures and then placed into long-term storage in facilities designed specifically for that purpose. Most often this involves freezing the remains and maintaining a very low temperature around them over time. Lack of this procedure will, as is the case with human remains, result in the eventual deterioration and decomposition of the bodies.

It is also a fact that alien spacecraft have, on occasion, been brought down intentionally by human military forces. Although confrontations between humans and aliens typically result in a one-sided victory for the alien forces, this is not always the case. There have been instances when craft of unknown origin have been intentionally shot down by human warplanes or other weaponry.

There have also been multiple instances of military radar interfering with the navigation systems of alien spacecraft, causing the pilots to lose control of the vehicles and resulting in the spacecraft crashing to Earth. As in the case of spacecraft which have crashed for unknown reasons, the occupants of these vehicles have been recovered by the military of various nations. Some have been deceased, apparently killed upon impact, while others have occasionally been found who survived the crash.

In all cases, both the dead bodies and the bodies of the survivors have been of a solid, three-dimensional nature when they were recovered. None were composed of ectoplasm. They had skin, organs, body tissue and most often had arms, legs and heads as well. Beings which are said to be of a demonic or angelic nature have none of these things, because their bodies do not take a physical form but are instead of a divine nature. They cannot be autopsied, frozen or stored. They cannot be injured by crashing to Earth, nor is there any reason they would require spacecraft to travel around in.

There seems to be little question that alien beings have been misinterpreted as divine or demonic entities throughout human history. This is a situation which may well have been happily accepted or even contrived by the aliens themselves.

The data is clear, however, about one indisputable fact: aliens with physical bodies are known to have been recovered and closely examined. This negates the proposition that they represent spiritual entities from what we commonly think of as heaven or hell. No matter how closely we may believe they resemble our concept of the appearance of angels and demons, it seems clear that these alien beings are mortal: they live, they age and eventually they die. They have a date of birth and they can be killed.

Regardless of what one's personal belief in religion may be, we cannot reasonably hope to understand either the aliens or their activities by equating them with divine beings of some kind. Such divine beings would have no conceivable use for such things as food, mineral resources, underground bases, weaponry, etc., but all these things and more have been associated with extra-terrestrial beings on numerous occasions.

Are certain alien beings able to inhabit and possess the bodies of humans in a manner consistent with ancient reports of demonic possession? Yes, they can do this—it has been reported by many reliable sources and I have personally seen it done.

Many people hold the belief that extra-terrestrials are really demons or angels, such as are described in The Bible and other places. This is a tricky subject to address, since people in general are often quite touchy about their personal religious beliefs. They can be quick to take offense when someone expresses an idea or attempts to provide information which contradicts—or appears to contradict—their already well-established beliefs about the subject.

It is difficult for many people to consider the alien contact phenomenon in an isolated fashion, rather than allowing it to become interpreted within the framework of whatever their personal religious beliefs may happen to be. This is something which must be avoided, in my opinion, at least initially.

While there is good reason to presume that alien beings have been identified as and confused with gods, demons, angels, spirits, etc. throughout the entire course of human history, it is a mistake to approach the subject in this manner. If we wish to gain a clear understanding of the entities we are dealing with, it is incumbent upon us to keep the topics of alien contact and religious belief in separate compartments as much as possible. Remember, the presumption that aliens literally represent demons or angels is nothing more than an opinion—it is not based upon the evidence.

This being the case, an unbiased researcher is required to reject any ideas which confer immortality or supernatural origins to these alien entities. To do otherwise will necessarily result in conclusions which are skewed and distorted by the introduction of improper assumptions and analysis. This will inevitably lead to an improper and incorrect understanding of these beings, as well as their nature and origin.

It is quite clear that alien beings, while often possessing abilities and characteristics which are quite different—and often far more sophisticated and advanced—than those possessed by humans, are neither timeless nor immortal. Under the right conditions, it is quite possible to damage the physical bodies of any alien beings which have manifested within our commonly-experienced dimension of reality.

Though it is highly unlikely that you will ever personally find yourself with the ability to carry out such actions, it is nevertheless true that it is possible to put a bullet right between the eyes of an alien being. If this is done, it will not be found that a bullet passes harmlessly through the body of a fully-manifested alien, as it would be expected to do if a gun were fired at some type of angelic or demonic entity.

Shoot an alien being between the eyes with a .45 magnum, and the result will be one very dead alien with a large hole through its head. We must be clear about this. There has never been a report of anyone killing a demon by shooting it in the head.

The dead bodies of alien beings have been recovered from the debris of crashed spacecraft on many occasions and have subsequently been autopsied by trained medical personnel. Their bodies have been cut and sliced open, their organs have been removed and carefully examined. They were not composed of ectoplasm, they were composed of solid matter and standard surgical instruments were successfully utilized to perform the autopsies. Such a thing has never been reported regarding a demonic or angelic being, and again we must be very clear about this.

Alien spacecraft have, on occasion, been shot down by human military weaponry, typically originating from fighter aircraft. They have also been brought down when military radar has interfered with and apparently disabled the navigational systems on certain alien craft. When either of these things happen, the debris is of a physical nature, though its exact composition may well include alloys and substances which are unfamiliar to us. Angels and demons do not require such craft and would presumably have no use for them.

Sometimes we encounter descriptions of craft like this in ancient historical manuscripts—Ezekiel's wheels, for example. The logical approach is to presume that what was being described was a physical spacecraft which was misinterpreted as a divine or supernatural device by people who lacked the sophistication to imagine them as anything else. This strongly indicates that what we are dealing with there is an instance of misidentification—it does not imply that angels were flying around in spacecraft.

I am not here to discuss religion per se. I have no interest here other than pointing out that for the purposes of the issues I am about to address it is imperative that no matter what a person's beliefs may be, we do have to agree on a definition of terms.

For our purposes, a belief which is based on faith—meaning a person is very sure something is true but has no physical evidence to support that belief—must be regarded here as opinion-based rather than fact-based. If there was physical evidence which supported such a belief it would not be defined as "faith".

Opinions can and do vary from person to person. When physical evidence is absent, however, it is impossible to objectively prove anything about anything. It is important, therefore, during a search for truth, to be clear about what can and what cannot be demonstrated or proven in an objective and convincing manner.

When discussing the subject of whether extra-terrestrials are literally the demons and angels described in various religions, we very quickly come to a basic issue which cannot be avoided. It is the question of the supernatural versus the natural. The issue can be boiled down to this: supernatural entities such as demons and angels exist eternally—they do not live and die. If aliens are such, they also can be presumed not to live or die.

This becomes rather important when we consider the fact that lots of stone-dead aliens have been found at the crash sites of alien spacecraft. If extra-terrestrials were in fact Biblical angels or demons, they could have been expected to fly away back home after such an incident with no permanent harm done to them.

The presence of burned and mutilated alien bodies scattered in with the debris makes that a difficult idea to defend in terms of objective logic. Actually, it makes it impossible. We cannot have angels or demons which go around leaving their dead bodies scattered in the wreckage of alien spacecraft. We can, however—and we do---find those bodies regularly.

Technically, this fact alone makes short work of the argument that extra-terrestrials are angels sent from heaven or demons sent from hell. In the world of objectivity and rationality, there is no choice but to conclude from this that aliens are physical beings which exist in a physical reality. They live, they die…and they can be killed. If you shoot an alien which is not wearing body armor dead center between the eyes, what you will have on your hands is one very dead alien.

That cannot be said of either demons or angels, though I highly doubt that anyone is likely to attempt to shoot a demon or an angel between the eyes to prove or disprove that idea.

If any of you happen to wish to attempt that and, after hitting the bull's eye, finds themselves face to face with an angel or a demon which is unharmed by the bullet, please try to convince it to accompany you to the nearest university for testing in a science lab. Otherwise, we are forced to reject that idea and accept the notion that we are dealing here not with supernatural entities but with physical beings which can sometimes appear very strange to us and very different than we are.

That does not mean, however, that we are finished with the subject, because it also seems to be the case that spiritual warfare is taking place between extra-terrestrials and humans and that this war is taking place not on the physical plane where our bodies exist but in realms of pure energy. When one refers to places such as the astral plane, it would not be improper to refer to them as spiritual planes, since everything there is composed of energy and we exist and perform actions there as an energetic—or spiritual—body rather than a physical body.

When an attack is carried out against our astral form, is it not in fact an attack against the soul? The soul, after all, can be said to exist as an entity of pure energy which is not bound by the crude matter of our physical shells. Isn't this also the definition of our astral bodies? Can we be imagined having more than one body composed of pure energy, one of which can be considered our "soul" and one of which cannot? If so, how would one tell them apart?

I do not think such a distinction can be made, especially in consideration of the memory I related earlier which involved the literal extraction of my soul from my body using alien-based technology. When our astral forms are attacked, it seems to me, our souls are attacked at the same time. Our essential energy, that which goes into forming and maintaining all our thoughts, memories, personality traits and talents, is being attacked by an external opponent whose intention is to inflict harm upon that astral form. If this does not constitute spiritual warfare, I am not sure what the proper definition of it would be.

I will at this point make a personal admission: after studying these things all my life, it is my opinion that we may well be better off if what we were facing were in fact angels and demons than extra-terrestrials. Extra-terrestrials appear to me to be a whole lot scarier.

Rather than facing entities from heaven or hell, we are instead facing enemies which exist in the physical world and are also capable of using the spiritual planes very effectively. They are intentionally attacking our astral forms—our souls—as well as our physical bodies for intentions which are purely their own and which to all appearances are clearly malevolent.

I would personally be much more comfortable with the idea that a god or devil I am familiar with is attacking my soul, rather than the idea that a strange non-human entity from another star system with abilities far superior to our own is waging war against the human soul for its own purposes. The idea that our government is surely aware of this fact and that we are told nothing about it even as the attacks are ongoing makes it even worse.

I do not pretend to have sufficient knowledge to analyze this situation in detail and have any hope of coming up with the correct answers. It is possible that someone else may have that knowledge, but if so the chances are good that they are employed directly by the government and sworn to secrecy at the point of a gun. In other words, we are probably not going to hear from them any time soon.

The idea that some—not all, but some—extra-terrestrials are described as inter-dimensional entities implies that they can and do interact with humans on planes of pure energy and to carry out operations on those planes. The fact that some of these inter-dimensional beings clearly appear to be self-serving, intrusive and hostile makes this an extremely dangerous situation, it seems to me, and one to which there appears to be no easy defense.

There is little question in my mind that some types of extra-terrestrials can inhabit and possess the human body and to utilize it for their own purposes when they do. It appears that something similar was done to me during an abduction episode. Certainly, I had no apparent contact with and no control over my physical body or what actions it may have been involved in carrying out at the time.

How can one tell the difference between a body which is possessed by an alien being and one which is being possessed by demons as is described in religious literature? I do not know the answer to that. I suspect that most or all the so-called demonic possessions in religious literature may instead be the result of alien control of a human body for purposes of its own.

What those purposes may have been, I cannot say…but I also can see no useful purpose at play if a human body is literally possessed by a supernatural demonic entity. What, after all, could it be hoping to accomplish? Is it anxious to have its clock cleaned by an exorcist?

Does it accomplish some overridingly important goal by making a human projectile vomit and float above a bed? I can't imagine what kind of advantage that would ultimately provide to a demonic entity.

At the same time, it is difficult to imagine what use a presumably highly-advanced extra-terrestrial life form would have for a human who could only speak in tongues and had an annoying tendency to engage in projectile vomiting. My final analysis of the purpose either type of entity would have for a human in projectile-vomiting mode is this: I have absolutely no idea and no useful hypothesis to put forward about it. So, let's move on, shall we?

Yes. Yes, we shall.

CHAPTER 6: WHITE NOISE

Be able to notice all the confusion between fact and opinion that appears in the news.
--Marilyn vos Savant

"Wots...uh, the deal?"
-- Pink Floyd

The Trouble with Pleiadean Ambassadors

Over the past several decades, I have met uncounted dozens of individuals who have claimed to be Pleiadean Ambassadors to humanity, or something similar. In addition, I have seen literally hundreds of these people in groups which cater to those kind of ideas, proudly proclaiming their alien heritage and announcing that they have come here—or been sent here—to assist the feeble-minded, spiritually-hindered humans on their path to ascendance.

Perhaps that is exactly what they are—I have no way of knowing for certain. I have yet to come across even one of those people, however, who was able to convince me that their claims were legitimate. Rather than being representatives of an allegedly-superior alien race, they do not appear to be superior in any way, as best I can determine.

They are not particularly intelligent—I would guess that some of them may be of average intelligence, at best, but most them do not appear to even qualify as average. It would seem to go without saying that someone who expects others to believe they are an Ambassador from the Pleiadean race, just because they said so, is unlikely to be a genius.

These people display none of the special abilities or powers which we might expect to find a Pleiadean Ambassador equipped with. They are not telepathic, they do not have the ability to materialize or dematerialize in our dimension, nor do they have any type of advanced technology to show us.

What they appear to be, rather than alien ambassadors, are egomaniacal underachievers who are desperate to be something special, something superior to those around them. This could be because they are unable to distinguish themselves among their fellow humans with their abilities, accomplishments or commendations.

They appear, for the most part, to be people who have become members of groups where people like themselves tend to congregate. These other members will most likely be inclined to accept their claims of official alien ancestry and status at face value, without being shown any type of compelling evidence which could substantiate them. There is no shortage of people like this—as P.T. Barnum once noted, there is one born every minute.

There is also no reason that I can think of which would make genuine Pleiadeans want to place their ambassadors to humanity among groups of people such as this. They obviously have no chance whatsoever to communicate any important messages the Pleiadeans may have given them to more than a tiny handful of relatively-uninfluential people, much less have them made available to humanity as whole. This would seem to defeat the purpose of any such activities and render it a rather pointless and poorly-conceived waste of valuable time (as well as presumably-valuable ambassadors).

Advanced beings such as the Pleiadeans and others, if it is their intention to give humanity one or more important messages, surely have far more efficient and credibly methods available to them to make their messages known. Nothing appears to be preventing them from making use of any of those alternate methods.

To believe that they would instead choose to utilize such an ineffective method of communication is something which defies common sense and logic. In my opinion, it should not be given the presumption of legitimacy by any thinking person. Rather than representing actual communications from advanced extra-terrestrial civilizations, it is almost surely nothing more than one more example of white noise, self-deception and disinformation which serves to confuse people and cloud the issues we hope to be able to understand accurately.

Anyone who would entrust messages to humanity to a small clique of fringe-dwellers who inherently lack credibility is, it seems to me, unlikely to be of much help to us. For that matter, they probably are not sufficiently-advanced to get here in the first place.

Perhaps, given their apparent proclivity for inefficient, poorly-conceived solutions, they would even now be safely inhabiting an airtight wooden dinghy of some kind, attempting to row their way to us across the vast, airless vacuum of interstellar space. If so, I think it is safe to presume they do not possess the answers we seek, or the level of planetary assistance we were hoping they might provide.

Please do not misunderstand my meaning here. I do not mean to imply that members of friendly alien races occasionally contact members of the human population. I can attest to the fact that this does in fact occur, based upon my own personal experience.

I can also state with certainty that at least some—and, quite possibly, *all*—of the alien beings who are involved in such contacts have no interest in having their selected contactees spread messages from them to humanity as whole. These contact events appear to be tailored to the individuals in question, for a variety of apparent purposes.

On some occasions, the contactee has been given knowledge of a particular subject by the aliens in question, often with the aid of advanced equipment which facilitates rapid learning. Other times, the reason for the meeting seems to have been little more than an opportunity to engage in a bit of friendly conversation between the two parties.

The point here is that such contacts and direct, face-to-face communications do in fact sometimes occur, for reasons known best to the aliens themselves. To the best of my knowledge, however, none of them have resulted in the human contactee being informed that they are now an officially-endorsed Pleiadean ambassador to humanity.

It is also a fact that direct communications between alien entities and members of the human population sometimes take place in what is known as the astral plane, or the dreamscape. Many alien races are known to be well-familiar with these energy-based realms, and to utilize them frequently and with a great deal of expertise. As with the direct physical contacts, however, such contacts do not end with the human being promoted to the rank of interplanetary goodwill ambassador.

It seems to me that, should an alien civilization wish to appoint official ambassadors to humanity, the individuals selected for such highly-esteemed positions would surely be people who have well-established reputations for honesty and credibility in the public eye. I would think that they would also be people who can avail themselves of widespread media coverage, to ensure that their responsibilities as ambassador are carried out in the most efficient way possible and that their messages are widely-dispersed.

If this occurs, it seems to me that it will be an event we will all certainly become familiar with in short order. We will surely not find it necessary to go searching for interplanetary ambassadors on the furthest fringes of the ufology spectrum, in discussion groups virtually no thinking person takes seriously.

Channeled Messages from Aliens

The most obvious—and problematic—issue here is that it is virtually impossible to produce any type of evidence that a conversation which supposedly resulted from channeling is legitimate. There is no way for anyone to be certain that such reported conversations were not invented wholesale for financial gain and notoriety by people who are, shall we say, less than honest.

Because of this indisputable fact, there is no reason whatsoever to presume that any of these supposedly channeled conversations occurred at all. It is far more sensible to believe that they are nothing more than fiction produced by greedy (and generally poorly-informed) hucksters who are out to make a buck. They do so at the expense of the good-hearted people who have chosen to take them at their word just because they happen to put on the appearance of a kindly old lady or a scantily-clad occultist of some kind.

I am not intending to imply that mind-to-mind communication is impossible, or that it never takes place. It is well-known that certain alien beings such as the Greys and others utilize telepathic abilities to communicate, and that humans are sometimes able to do likewise to a certain extent.

It is not an ability which most of us possess to any great degree—at least, not as far as we are aware. But the mere fact that communication with the Greys takes place on an ongoing basis with certain people who are attached to the military serves to demonstrate that mind-to-mind communication is indeed possible and is being utilized in specific instances. To believe without evidence that a typical "channeled conversation" between a random member of the public and an ambassador from an alien nation, however, is a very different thing and something which, it seems, makes very little sense to do.

This idea is given even more support by the second issue regarding channeling, which is its inherent inaccuracy. The records of purported channeled communications are rife with predictions of certain events—often very specific ones—which are attached to specific dates. To the best of my knowledge, none of these predictions have actually come to pass at the correct time which it was predicted—sometimes even guaranteed—that they would take place.

Again, this is a major problem when it comes to accepting channeled conversations as being legitimate. Surely no ambassador of an alien race would fill the channeled records with specific predictions which did not then occur. To do such a thing would clearly serve to greatly decrease the credibility of both parties involved in the communication.

It is fair to say that even a comparatively less intelligent being such as a human would not have to think too long or too hard to understand the potential problems making such unfulfilled predictions would cause in terms of having their messages be given the presumption of legitimacy. Enough misses and these messages would, it seems, be far more likely to be rejected as a matter of course than accepted as authentic.

It should be noted that some of these supposedly-channeled messages are said to have come from Grey aliens, who proceed to describe their home planet and way of life. It is, according to some of these messages, a place of spiritual bliss and enlightenment. Anyone who is aware of the origin and the nature of the Grey aliens will immediately know that such a claim is manifestly untrue, and that anyone who makes it is either being lied to by a malevolent alien being who wishes to manipulate them or is engaged in outright fraud and should be ostracized from the community of serious researchers.

Another issue with many of the allegedly-channeled messages is that they often contain assurances that friendly alien beings have seen fit to send a war fleet to assist us in ridding ourselves of the hostile groups which reside here. Some even claim that this is already being done.

Unfortunately for us, these supposed warships are conveniently invisible, and any indications that we are receiving assistance against hostile aliens are conspicuously absent. Nobody from the inside is filing reports about "blue avians" either. Invisible war fleets rendering unnoticeable aid and is a rather large problem in terms of convincing me to take reports such as these seriously and, in my opinion, provide a very good reason not to.

A third problematic issue which is attached to the idea of these supposedly-channeled messages and declarations is that of the selection of the individuals who claim to receive these communications. Do they represent the best and brightest members of our society? Do they include top physicists, political leaders and university professors?

To the best of my knowledge, people such as these are never among those who are supposedly specifically selected by representatives of an alien race to be the vehicles to which their messages to humanity will be transmitted and entrusted. They are, rather, people who seem to possess no special qualifications at all in terms of being reliable, competent, trustworthy or capable of passing the messages along to a worldwide audience in a manner which will be taken seriously.

That is a major mistake. It is one that even I—a relatively dull-witted member of a backward race—would not even come close to making, were I in charge of such a project. I highly doubt that any of my readers would, either. It seems extremely unlikely, therefore, that a race such as the Pleiadeans would choose to resort to little old ladies from Lithuania or middle-aged wallflowers struggling to make ends meet on their Social Security allowances to be on the receiving end of their channeled transmissions. *We,* after all, are the idiots in this conversation—not *them.*

Some people will surely protest, claiming that the supposed recipients of these communications are chosen on the bases of their superior levels of "spiritual enlightenment" or whatnot. This is, in my opinion, nothing but nonsense. It is a claim which cannot be substantiated and which there is no reason to believe is accurate. People who claim to receive channeled communications do not appear to demonstrate any discernible superiority over others in terms of their spiritual enlightenment, other than their own assertions that it exists.

It should also be noted that there have been many individuals who are widely accepted as possessing spiritually-advanced souls— Mahatma Gandhi and the Dalai Lama are a couple prominent examples—who were apparently passed over and never contacted mentally by members of an alien race, even though they would appear to be obvious and highly-visible choices to make.

As best I can determine, there is no way of getting around this problem and no way of assessing it which accomplishes anything in terms of adding to the credibility of most channeled messages. From the point of view of those who would have us take such communications seriously, it would seem to be all bad. And we are not quite finished making note of the problems which are inherent to many such claims.

We come now to the question of why beings which claim to represent several non-human races of off-world origin would choose to utilize a method such as channeling in the first place. If it is their intention to cause the residents of planet Earth to become aware of some type of essential information, channeling as it is portrayed in the New Age school of belief would be perhaps the least efficient means of accomplishing that.

The same principle can be applied to the idea that a race of enlightened, friendly alien beings would choose to send a message of hope to the people of Earth by making personal contact with a few apparently random, seemingly-unqualified people. These people are then entrusted to pass along this message to the rest of humanity, and to do so in a manner which the remainder of humanity would find both highly-credible and compelling.

It requires no more than a brief look at the world around us to observe that no such thing is taking place, nor is there any reason to believe that it will in the foreseeable future. Most of humanity goes about their daily business completely unaware of any purported messages of hope which originate from the Pleiadean race. They are utterly unconcerned about the beliefs of those involved in the New Age movement and generally consider most or all the members of that group to be rather soft in the head.

Although it is true that both these types of communication are accepted by many individuals within the New Age community, in terms of the population of the world at large that number is quite miniscule. Considering that the ideas espoused by the New Age community in general are for the most part either unknown to or completely disregarded by those outside their group, the chances that these messages would become known to and accepted by the rest of the world are virtually nonexistent.

This leads to the question of why any advanced race would choose to make use of either of these methods to impart a message of any significance whatsoever to the people of Earth. They clearly would be able to utilize global radio or television transmissions for such a purpose, in which case their goal would be accomplished virtually instantaneously, with people worldwide all receiving the message at the same time. Hack into the worldwide television network, deliver the message personally, problem solved.

Failing that, even the idea of flying over our cities in their ultra-advanced space vehicles and dropping several billion leaflets to the inhabitants of Earth as they do would be a vast improvement in terms of mass communication than the methods they are currently said to employ. It doesn't require the mind of a super-intelligent alien being to come up with this rather obvious idea. Why, then, is it not being done?

There are several possible reasons for this, none of which bode well for those who espouse the principles of New Age thought. The first is that they are unable to accomplish this for some reason. Clearly such a reason would not be based on some technical inability—any alien race which can get here in the first place must surely be presumed to operate at a level of technical sophistication which would far surpass our feeble abilities. Hacking into a worldwide television or radio grid would be, as far as they are concerned, child's play.

If, then, a race of alien benefactors is unable to contact the people of Earth in an efficient, indisputable way, the reason for this could be because they are being prevented from doing so by some outside group. Perhaps, one might reason, they have been prevented from doing so because of some type of secret agreement they have made with one or more Earthly governments.

Of the list of possible suspects, the government of the United States of America stands as both the most prominent and the most likely possibility. Due to America's current position as the world's only remaining superpower, the idea of an alien race being so constrained due to having made agreements with other world powers such as England, Russia, China or others does not contain much in the way of credibility. If America wanted the world to hear some Pleiadean message of hope and salvation, any agreements that the Pleiadeans may have signed with those other nations would surely be insufficient to prevent the word from going out.

Whether it is fair in the view of others or not, the reality of the current state of global affairs—both now and for most of the past century—can be summed up as "What America wants, America gets". That applies to finance, trade, military power and cultural influence—it goes without saying that it would certainly apply in the case of a message from alien Space Brothers that the government of the United States was intent on passing along to the citizens of the rest of the world.

It is known to me that the United States does indeed both require and abide by a non-disclosure agreement contained within the body of the Greada Treaty. Non-disclosure was something both parties to the treaty insisted on, in fact, according to my sources.

Public disclosure of the facts would make things more difficult for both signatories of the agreement. It is, after all, far more difficult to abduct humans—some of whom you decide to consume, or to never return home for another reason—when the entire world knows what you're up to and has started staying awake all night, armed with assault rifles and guarded by teams of house cats (Grey aliens *detest* house cats for some reason).

Similarly, it is far more difficult to run the military if the population of the entire planet knows that the aliens have landed, that the news is very bad and that the most powerful military juggernaut in the history of the world is defenseless against them. It is also far more difficult to get away with kidnapping people to the underground levels of Dulce Base and keeping them there for life (where they will be subjected to horrifying, illegal and unthinkably immoral genetic experimentation) if the population is awake and armed.

But the Greada Treaty has nothing to do with this. That agreement involved alien beings far different from the Pleiadeans, and beings who by all appearances have already wrested control of this planet from the humans and installed everything necessary to promote their imperial agenda.

Whether the secret government of the United States also made an agreement with one or more alien races, races who are not party to the Greada Treaty, which also contains a mutual non-disclosure clause is a matter for conjecture. It has been reliably reported that those in the upper echelons of the American power structure have indeed signed at least two additional agreements with other alien races. The specific alien races/groups which may be party to any such treaties and the terms these agreements may contain are unknown to me.

If non-disclosure was, however, a matter which was stipulated in the language of an agreement with, for example, the Pleiadeans on behalf of their entire race, I presume that it would certainly include a ban on such things as channeling messages to the citizens of the planet and sometimes contacting them personally in the physical world. I can tell you for certain that it would, if I were the one being asked to sign such a document on behalf of the secret government of the United States.

All things considered, it seems unlikely to me that the Pleiadean race is party to a non-disclosure agreement with the American government. Whether individual groups of Pleiadeans may have some such arrangement, it hardly matters in terms of our present discussion. This leads to yet another possibility: certain groups of friendly alien beings may wish to send us a message but are being prevented from doing so by one or more third parties. In other words: they want to, but somebody won't let them.

This leads to the question "So…what's up with that, Pleiadeans? Are you lying to the President and the military, so you can spread your message of honesty, goodness and light? If you are indeed capable of channeling these endless streams of mindwash drivel to scores of gullible idiots around the world, surely, it's not too much bother to channel the answers to such simple and sensible questions directly to me.

You may proceed with the transmission at will. I await your non-physical presence. Have at it, boys.

(Note to any Pleiadeans who may happen to be perusing this manuscript: I suppose that by now you may well be tempted to internally horse-whip me with some type of horrible scalar weapons system or something. If so, please take note of the fact that you are, at best, third in line and may well be even further down the list than that.

Also, please take note of the fact that it's highly unlikely I'll survive if either of the first two names on the list decide to make use of their option, and you'll be out of luck. So...talk to me about your concerns later, I guess, because right now I am busy erecting a wall of concrete-reinforced sandbags around the perimeter of my survival bunker.)

Tales of Mighty Bigfoot

Within the community of ufology buffs, there exists a considerable number of people who are quite convinced that the creature commonly referred to as "Bigfoot" is far more than it might at first appear to be. According to these people, Bigfoot is a highly-intelligent, telepathic entity which has utilizes teleportation portals to facilitate long-distance transportation for itself as necessary. Some feel that Bigfoot is a native of our world—though perhaps not of our dimension—while others are convinced that Bigfoot is of alien origin and has travelled here from its original home far away.

Now, I don't know where you might happen to be from, but I will tell you that I am from a place which must be described as "Bigfoot country". Both Roger Patterson and Bob Gimlin, who were responsible for the most famous (and controversial) film of a creature which appeared to be

Bigfoot, clear back in 1972, were life-long residents of the same town I grew up in.

Though I cannot claim that I knew Patterson personally, I did meet him briefly a couple of times when I was high-school age. I did not have an opportunity to ask him anything at all about his sighting, nor did he mention it.

I did have the opportunity to become acquainted with Bob Gimlin many years later, when I was in my mid-thirties. Again, I will not claim that we were close friends, but our paths crossed several times and he always conducted himself in a polite and friendly manner when I spoke with him.

Unlike was the case with Roger Patterson, I did have the opportunity to ask Bob about his famous sighting once, during a casual conversation which took place at a horse ranch which was owned by a mutual friend of ours. I asked him, quite naturally, if the famous sighting and film were legitimate, or if they had been the result of some type of monkey business.

This is what he said to me on that day: "Derek, I don't know what the hell that thing was. I've never seen anything like it in my life, before or since. If that's what a Bigfoot looks like, then we saw a Bigfoot. But I don't know what the hell it was."

Gimlin had no reason to lie to me about it, and I believe that what he told me was the truth. He was constantly asked about the subject, of course, and soon became sick of being pestered about it all the time, often by complete strangers. He sold his entire interest in a movie which was made about the incident for the price of one dollar, hoping that doing so would result in people leaving him alone and letting him have some peace and quiet.

Patterson, who passed away many years ago, insisted that both the sighting and film were the real thing, from the day the incident took place for the entire rest of his life. He went to his grave swearing that he and Bob Gimlin had, in fact, seen and filmed a Bigfoot that day, in a northern California forest.

I have never personally seen anything which resembled a Bigfoot and have no way of determining with certainty whether the film clip is, in fact, legitimate. It should be noted, however, that although the film has currently been around for many decades. It has been carefully analyzed by numerous people of all types over the years, and it has never been proven to be fraudulent or successfully debunked by anyone.

I have no reason to believe that either Patterson or Gimlin was particularly motivated to lie about something like that, and it would seem rather pointless to engage in fraud simply for the sake of a ten-second piece of amateur video. I choose, therefore, to take them at their words as far as the film clip and the incident itself are concerned, because I have no good reason not to do so.

Does a creature called Bigfoot truly exist, living deep in the old-growth forests of the Pacific Northwest (and, reportedly, several other locations as well)? I am the first to say that I have no way of knowing for sure, one way or the other.

I will add, however, that Patterson and Gimlin are not the only people I've met over the years who claim to have sighted Bigfoot at some point. Several other people I know, some of whom I consider to be quite credible, claim to have seen Bigfoot at least once within the vast stretches of wilderness which comprise the Cascade Mountains.

To the best of my knowledge, there has never been a credible Bigfoot sighting which involved the creatures appearing from, or traveling through, any type of space-time portal. Furthermore, there is no evidence anywhere that Bigfoot is some type of alien being which has traveled here to Earth in an interstellar spacecraft.

All reports which relate either of those things to Bigfoot clearly appear to be pure invention, and many of them almost certainly originate from professional disinformation personnel. It is nothing but one more wild goose chase, a way to keep the attention and time of others diverted and safely occupied with things that will ultimately make no difference at all, because they are not even true.

Some psychics have reported achieving telepathic contact with Bigfoot. If one were to make the presumption that every so-called "psychic" who claimed such a thing was engaged in fraud and lying about their claim, I feel very certain that one would never be proven wrong by doing so.

It should also be pointed out that none of the inside sources who are widely considered to be honest, legitimate and credible, have ever made any mention of an alien race known as the Bigfoot, or of space/time portals being utilized by Bigfoot as a method of travel. If such things existed here in the real world, we surely would have been told about it by now, through a reputable and reliable inside source.

Since that has not occurred, we must presume that there is a reason for that. The reason is not that all of them simply forgot to mention it, it is that those things do not exist in the real world.

Gigantic creatures do not travel here through portals, on missions to wander naked through the forests searching for berries and accomplishing nothing else, as best we can tell. The idea that they do is sheer nonsense, and those who believe it to be true are believing in fairy tales and fraud.

I will never forget a post I saw in a Facebook group once, regarding Bigfoot. A guy had posted the following: "I received a telepathic message last night from Proud, Noble Bigfoot. All he wants is Peace".

Well, Bigfoot Guy, if you happen to be reading this: another message has come for you, this time from a human. Someone—let's call him Helpful, Observant Derek—wishes to inform you that you are an idiot, and that future generations would be well-served if you are prevented from reproducing by any means necessary.

I'm sure there are some of you who want Bigfoot Guy to be telling the truth. I can tell you with great confidence, however, that—when it all comes down—the smart money will be betting that my message to Bigfoot Guy was the accurate one, and that the telepathic communication from "Proud, Noble Bigfoot" was nothing but juvenile, completely fabricated, nonsense.

Alien contact is an incredibly complicated, extremely broad topic, under the best of circumstances. It is crucially important that we find some way to achieve a basic understanding of the situation as it exists in real life. Nothing—and nobody—is helped by wasting time on fantastical, pointless, unprovable nonsense which does not improve our understanding of the subject but wastes our valuable time instead.

If you happen to be one of those who has been engaged in the hot pursuit of a portal-using Bigfoot who sends telepathic messages to humans for no apparent reason…please, do yourselves an enormous favor and give up the chase. Move on to something which is real, and which can help you to increase your understanding in ways which will work to your advantage.

CHAPTER 7: THE SERPO DOCUMENTS

"I have a higher and grander standard of principle than George Washington. He could not lie; I can, but I won't."
— *Mark Twain*

"Tell me lies, tell me sweet little lies,
Tell me lies!"
--Fleetwood Mac

Over the past several years, a collection of documents has been published on the internet which are alleged to contain first-hand accounts and testimony pertaining to something which is referred to as "Project Serpo". According to these documents, Project Serpo was the designation which was given to a highly-secret agreement between the United States government and that of the race of alien Greys which were associated with the notorious Roswell Incident in 1947.

According to this agreement, it is claimed, both parties agreed to participate in a mutual exchange of personnel, which would remain in effect for a designated period of years. According to the information contained in the "Project Serpo" documents, a group of human ambassadors would, at a pre-arranged time, be dispatched to board a spacecraft which belonged to the Greys. They would then be transported safely to a planet called Serpo, in the Zeta Reticuli star system, which was said to be the home world of these Greys.

Upon arriving at Serpo, the contingent of humans would take up residence there as official guests of the Greys. They would be allowed to interact and communicate with the inhabitants of Serpo. They would also observe and record as much as they were able to, to gain an understanding of the society, culture, beliefs and habits of these Grey aliens.

While this was happening, a contingent of the Greys would be tasked to remain on Earth and perform the same function, observing and learning as much as they could about human society and practices. The purpose of this alleged exchange of ambassadors was said to be an attempt to promote friendship, trust and understanding between the inhabitants of both worlds.

There have been rumors for many years in ufology circles that such an exchange of personnel did in fact take place at some point in the early 1960's, and that it was to a large degree considered to be a successful undertaking. Many people within the ufology community believe these stories to be based on true accounts. They believe that an exchange of personnel did indeed take place, and that the information contained within the documents which purports to recount the details of Project Serpo is largely accurate.

I do not share this view, for several reasons. Firstly, I see nothing to be gained as far as the Grey aliens are concerned from such an exchange of personnel. They have clearly been visiting our world for a very long time and are already well-familiar with humanity, our culture, our technological abilities and our military assets. They have had the benefit of long years of observation and interaction with humanity and know us extremely well by now.

Additionally, the Greys apparently have worked in cooperation with the American government for a most of a century. They already have personnel stationed in a variety of bases and facilities, both on Earth and in other locations in our solar system, almost certainly including the so-called "dark side" of the moon.

Considering all the knowledge about humanity they already possess, the idea of leaving a team of observers here to learn about us makes no sense. Doing so would be redundant, nothing more than a pointless waste of resources. What could they possibly hope to learn about us from this that they do not already know?

Certainly, we might wish to send observers to their home planet. As best I can tell, however, the Greys would have no reason to want to allow a team of human military personnel to travel to their home world, observe them closely, and report their findings back to Earth.

Giving away that kind of information, while receiving nothing substantial in return for doing so, would constitute a tactical mistake that it is hard for me to imagine they would make. Allowing us to closely observe their culture first-hand, including the use of photography and film to record it for posterity, could certainly be to the Greys disadvantage. There are no circumstances, however, where it could bring them any type of useful benefits.

To do such a thing would be much like inviting a team of Chinese spies, equipped with the finest video cameras available to them, to take up residence at NORAD headquarters for several years then report back to their leaders. I would not allow it to be done, and I highly doubt that whoever oversees such things for them would allow it either.

In addition to the tactical difficulties such an exchange would entail, there are many things contained within the Project Serpo documents which either do not make sense, or do not ring true to me. I will briefly discuss a few of them here, for your consideration.

One aspect of the report that I did not find to be believable was its descriptions of the Greys they supposedly interacted with during their stay on the planet Serpo. According to the Serpo Documents, these beings were very kind, helpful, eager to please and assisted them in any way they could. Even though a considerable number of reliable inside sources with first-hand experience to draw on have chosen to publicly speak about this alien race, not even one of them has said any of those things about them, ever.

Instead, the descriptions are generally in agreement with each other regarding the temperament and attitude of the Greys toward humans. They were reported to have little regard for the lives of individual humans and demonstrated a distinct lack of concern or remorse when humans were killed because of their activities.

There have been numerous reported incidents which involved Grey aliens killing humans, usually either military personnel or laboratory workers within bases which were jointly occupied by both humans and Greys. The Greys are sometimes said to have been reprimanded for doing so by the top-ranking humans at the facility. Aside from that, they were never actually punished for having killed any of the people who were assigned to work with them at the base.

In addition, I am told that the terms of the Greada Treaty allow them to terminate any life-forms, including individual humans, which are found to be "genetically inferior" or "otherwise unacceptable". I have no way of knowing how many people may have been judged to be genetically "unsuitable" to be allowed to continue living. The mere fact, however, that they specifically sought—and received—such permission implies that it must surely be a sizeable number, over the decades that the treaty has been in existence.

This is a far cry from beings which are helpful, friendly and kind to humans. It is in direct contradiction to virtually all testimony about the Greys which has come from reliable inside sources. Even if there were no other problems to be found within the Serpo Documents, this alone would be enough reason to consider them to be disinformation rather than an accurate accounting of true events.

According to at least one inside source, the Greys do not hold a belief in an all-seeing, omniscient deity as many humans traditionally do. Instead, they believe that a universal life force or universal consciousness is what matters, and that the lives of individuals are not of much concern in the overall scheme of things. They regard the taking of human lives, according to this person, in much the same we think of popping balloons, and therefore not something they need to be either particularly concerned about or careful to avoid.

This is in direct contradiction to the information contained within the Serpo documents, which reported that the Greys gather together each evening to perform a group prayer ceremony which honors whatever gods they happen to believe in. It seems to me that this is nothing but pure nonsense. In my opinion, it is something which was almost certainly inserted into the report to make the Greys appear to be more sympathetic in our eyes, and more like us than they really are. It is surely nothing but pure disinformation.

The inclusion of intentionally deceptive material in these reports in this instance, as with the one discussed previously and others, gives us a legitimate reason to dismiss the entire contents of the documents on a wholesale basis. Lies in one place directly imply that lies are present in other places as well. It is, therefore, quite safe to presume that the whole thing is nothing more than white noise and confusion, intended to lead us into a blind alley and prevent us from using our time more wisely. I see no reason at all, however, to believe the Serpo Documents are authentic, or that the events they allegedly refer to did in fact occur.

We should also take note of the fact that, were the Serpo Documents authentic, they would certainly have been removed from the internet as soon as they were posted there. Instead, they have been allowed to remain in public view for years without being disturbed.

I cannot imagine that this would be the case if documents of this type were legitimate and appeared on a website which is dedicated to publicizing them. If they were real, they surely would have disappeared long ago, presumably along with both the webmaster and whoever happened to release them in the first place.

I feel it is necessary, before making the following statements, that I make note of my inherent limitations. I must carefully avoid any temptation I may feel to over-estimate my importance or qualifications within the universal command structure and inter-galactic food chain. Allow me to freely admit, therefore, that I am nothing more than a member of a dull-witted, primitive, socially-backward species of barbaric, uncouth ruffians.

Considering these serious disadvantages, I am clearly in no position to make judgments concerning what the leaders of a highly-advanced, intellectually-superior race of alien beings might think or do. All I can say is that, were I in charge of a planetary civilization whose existence was threatened, or which faced other serious problems, I would not allow my citizens to waste their time playing communal games much like volleyball every evening (yes, they were reported to do this).

Neither would I assign them to assist and educate a group of strangers from another world. I would surely be able to find far more important tasks and responsibilities for them to attend to. But, according to the Serpo documents, they played their little game every night, as though they were all members of a freshman P.E. class.

The final issue I will mention here is that, according to the documents, two members of the human team decided they wanted to remain on Serpo, rather than returning to Earth with the rest of the team. They sought, and received, permission to do so, according to the report.

Why would anyone in their right mind do such a thing? What were they planning to do, spend the remainder of their lives on an alien world, eating tasteless grey mush and playing volleyball in the evenings? Attempting to find love with a flat-chested, bald-headed alien female, while never setting eyes on either Earth or another human again? I am certain I do not know of any human who would voluntarily make such a choice.

I also do not believe that the officer in charge of the group would have been given the authority to grant permission to remain on the home planet of the Grey aliens to any of the members of the team. Doing so would defeat the purpose of the mission, which was allegedly to study the Greys and then report their findings back to Earth. If certain members could choose to remain there instead of returning to Earth, they may as well have never been included on the team in the first place.

Not only that, what was to prevent the entire team from doing likewise? I cannot imagine permission for such a thing either to be sought by any sane, rational human, or to be granted if it were, or for the commanding officer to have been given the authority to allow it in any case.

The idea simply makes no logical sense. In my opinion, it is another example of something contained in the Serpo Documents which is not—and cannot—be true.

There are more things which could be listed, as well, but I think this is more than enough to demonstrate that the Serpo Documents are nothing more than a poorly-executed disinformation operation. It was an attempt to deceive the public by supplying details of an expedition which did not actually exist in the real world.

To those who have not yet taken the time to read these documents, I would recommend that you do not bother doing so. The documents are extensive and take a considerable amount of time to get through. When you have finished reading them, you will find that you are several hours closer to death, with nothing of value to show for it.

Wasted time is one of our greatest enemies and reading the Serpo Documents will surely result in plenty of that. Do yourselves a favor and skip them.

CHAPTER 8: SHUTTING DOWN THE SILOS

*"We lost between 16-18 ICBMs at the same time UFOs were in the
area. We were ordered to stop the investigation, to do no more
on this not to write a final report. I heard that many of the guards
that reported the incident were sent off to Vietnam."*
--Captain Robert Salas, USAF

*"We'll know for the first time
If we're evil or divine,
We're the last in line!"*
-- Ronnie James Dio

There is a common belief going around these days which says that
aliens saved the world from a devastating nuclear war. They achieved
this by shutting down the ability of numerous silos to launch their
multiple warhead-equipped ICBM's at enemy targets, and thereby
usher in Doomsday, a thermonuclear exchange which had the
potential to destroy every living thing on Earth. I would like to
address this belief in this chapter, and to examine it closely to
determine whether it is factually correct.

If it turns out to be factually correct, we owe the aliens an
inestimable debt of gratitude, the scale of which makes it impossible
for us to ever repay. If, on the other hand, it is *not* correct, we need to
determine exactly why that is, and why so many people currently
believe it to be the case.

As it happens, determining whether it is indeed true that alien
beings prevented a devastating nuclear holocaust is a relatively
straightforward affair, and the conclusions it allows us to reach are, in
my opinion, both compelling and difficult to reasonably dispute.

Is it true that alien spacecraft have interfered with the launch
systems and the electrical power grid which operates them, at our
nuclear-capable ICBM silos on multiple occasions?

It is indeed true. It is also a fact that these instances of third-party
interference with our intercontinental ballistic missile launch control
systems was done without any prior notice or warning.

On March 16th, 1967, in Central Montana, multiple personnel assigned to an ICBM installation reported sighting a red, glowing UFO which approached the silos and hovered over them. As it did so, the guidance and control systems of United States' intercontinental ballistic missiles were interfered with by the unidentified craft. One by one, the guidance systems went offline, and the missiles became inoperable. In total, ten missiles were taken out of commission. Control of the missiles was not re-established until a full day later.

A short distance away, at another silo, an additional eight missiles were taken offline in a comparable manner, as a UFO hovered over and around the installation. A total of eighteen missiles were rendered inoperable that day by unexpected, unwelcome interference from UFO's.

This has been reported by Robert Salas, who was the Deputy Crew Commander on duty at the time. I have spoken with Mr. Salas, and I will tell you that he is a very serious, non-nonsense man who has no patience for foolishness or speculation about things which cannot be proven. It has also been verified by official USAF documents which were released later.

These were not the only installations which were interfered with in this manner. As time went on, numerous other missile silos were neutralized in the same fashion while alien spacecraft were present at their locations.

The identity of the occupants of the alien craft which were involved in shutting down the power grids of these most critically important, indescribably dangerous and heavily-secured installations was completely unknown to—and unobtainable to—the human personnel who were on duty at the silos in question. They had no way of determining precisely who it was that had chosen to shut down their entire electrical grid, and no method of determining what the reason for such an action might be. They were utterly unable to determine whether the aliens in question were considered friendly or represented a hostile alien civilization.

This created an extremely unpleasant situation for the humans who were on duty in those silos at the time, to say nothing of their superiors at the Pentagon. When one controls thermonuclear missiles with intercontinental range, and suddenly loses all power to the facility, one would certainly want to be able to determine who had caused that situation to occur and why they had done so.

Without being given any advance knowledge of the planned shutdown, and considering the incredible destructive power of the ICBMs, the lack of the ability to control these weapons required them to presume that the aliens who were responsible for the problem were—at a minimum—not our friends.

In the worst case, they were hostile and invasive beings from another world who had just neutralized our most powerful offensive weaponry. Without certain knowledge one way or the other, the only sensible and responsible thing to do was to consider the aliens to be hostile, and the shutdown of the power grid to be an aggressive, intrusive, uncalled-for and insanely dangerous action.

Technically, it was also an act of war and a direct, unprovoked assault on one of America's most sensitive and highly restricted military facilities.

After the incidents were over, and control of the silos in question had been restored to their crews, no explanation of any kind was forthcoming from the occupants of the alien craft which had clearly been responsible for the interference. No admissions of responsibility, no lectures, no dire warnings, no official transmissions, no communication of any kind was received from the offending craft.

Additionally, at least one incident occurred which involved a silo owned by the Soviet Union, and which was much more alarming than those which took place in America. During that incident, the launch timers of the entire complement of missiles housed in the Russian silo were remotely, and without warning, activated by the occupants of an alien spacecraft.

The occupants of the silo in question suddenly found themselves unable to delay or cancel the imminent launch of multiple thermonuclear missiles which were targeted at various American cities. The launch timers had been activated without warning by alien intruders, for an unknown purpose, and were now counting down the seconds until a launch that would surely result in nuclear annihilation for most or all of humanity would occur.

At last, with only seconds remaining before the countdown timers reached zero and the launch was initiated, control of the missiles was restored to the Soviet crews inside the silos, and the launch sequence was immediately aborted. Those seconds were literally all that stood between humanity and the end of the world as we know it.

When it comes to determining whether humanity was prevented from destroying itself by the actions of friendly aliens, it is a straightforward matter. Firstly, we need to determine whether, when these interferences with our launch capabilities and control systems took place, the world was on the verge of all-out nuclear war. We need to find out whether the troops in those silos, who were locked into their control seats by handcuffs they were unable to open themselves, were preparing to launch their deadly cargos as the first strike of what would likely have been humanity's final war.

This was clearly not the case when any of the missile solos were interfered with by alien intruders. Apart from the Cuban Missile Crisis in 1961, there has never been a time when the world was teetering on the brink of nuclear war, or when the onset of a nuclear exchange between the United States and Russia, or any other nation, was imminent. Humanity has many faults, but the mass suicide of the entire human race is not an option which would even be briefly contemplated by any sane, rational national leader or military commander.

In other words, alien beings interfered with the power grid, control system and launch capability of intercontinental ballistic missiles with multiple independent thermonuclear warheads attached to each individual missile, at a time when nobody had even the slightest intention of launching those missiles, and in addition had no conceivable reason to do so.

It is also a fact that a nuclear first-strike by either the United States or Russia cannot be prevented simply by neutralizing a couple of missile silos. Each nation has numerous silos in widely-dispersed locations, most or all of which would presumably launch their missiles at the same time in the event of a planned first-strike attack.

In the case of the United States, which has over ten thousand nuclear weapons in its arsenal, preventing the launch of twenty of them will make no difference at all, in practical terms. Russia, while having fewer total warheads at its disposal, is still in a comparable situation: neutralizing one silo will in no way prevent them from carrying out a first-strike attack, or launching waves of missiles in response to one, should they choose to do either of those things.

This does not even consider the fact that both nations possess multiple "boomers"—ballistic missile submarines which are armed with nuclear warheads. Each of these submarines is fully capable of inflicting devastating harm on any enemy nation in the world, all by itself, and of doing so within a matter of minutes, probably from a location only a few miles offshore of the nation in question. Submarines alone possess far more destructive power than is necessary to completely cripple and destroy any earthly opponent and operate so silently that they cannot be located or tracked once they slide beneath the surface of the ocean.

Additionally, land-based airplanes and warships of both the USA and the USSR have the capability of launching nuclear weapons if ordered to do so. American bombers based in Alaska, just fifty miles from Russian territory, can be airborne within a minute of receiving the orders. The nuclear capabilities of legitimate global superpowers are not limited to just one or two viable options.

So…did the aliens prevent a nuclear holocaust by shutting down our missile silos? Absolutely not. What they did was interfere aggressively with critically important components of our national security apparatus and military arsenal. They did so without warning or explanation of any kind, in a manner which was distinctly threatening.

These incidents had nothing to do with saving humanity from itself. They were nothing less than a power play on the part of the aliens. They were making it clear to both the United States and the Soviet Union that they had the ability to both initiate a thermonuclear holocaust at any time of their choosing, and to prevent either side from being able to defend itself from a nuclear attack. The unstated presumption was that the aliens possessed a sufficient number of craft to activate the launch sequences of all the necessary silos, or disable them, should they wish to do so. It is difficult to imagine a more threatening display of power than that.

Those who believe that aliens prevented us from fighting a third world war have not analyzed the situation adequately or drawn the correct conclusions about it. The alien interference with ballistic missile silos had absolutely nothing to do with preventing an imminent missile launch. It had everything to do with demonstrating their technological superiority in a way which was guaranteed to strike terror into the military and political leadership of the world's superpowers. It was a blatant act of war against both the United States and the Soviet Union. It was carried out in a fashion which made it impossible for either of the two nations to respond aggressively to.

The correct analysis of these events is, in my view, indisputable. The aliens who were responsible for these events—and, though I can't know for certain which that was, it is logical to presume it was the Greys—are obviously not our friends. They have demonstrated in no uncertain terms that they are not to be trusted, and that their intentions here have nothing to do with our best interests. Rather, they seem to be concerned with establishing and maintaining control over earthly governments and military institutions, while advancing their own agenda by any means necessary.

To presume that this agenda is anything other than hostile would be a foolish—and perhaps ultimately fatal—mistake on our part. From the preponderance of the evidence, it seems to be the case that the American government, and perhaps others as well, have secretly made a deal with the devil and, in doing so, have effectively sold out the entire human race.

The government is motivated to manipulate the public into either believing that aliens do not exist or, failing that, to convince them not to worry too much about the intentions of those aliens. We, however, do not have to be victimized by such self-serving tactics. Though it is true that friendly alien visitors certainly do exist, it is difficult to escape concluding that it is alien beings of a distinctly hostile nature who have established control here.

Judging by the situation we find ourselves in, it appears that the so-called "friendly" aliens are either unable to prevent the "bad guys" from carrying out their agenda here or, if they can, are unwilling to do so for some reason. Whichever of those things is true, it does not bode well for us.

Though the dangers we face may not be apparent to most and are never mentioned in the mass media, they are very real and critically important to our long-term survival prospects. We need to be quite candid about that. Kidding ourselves into thinking otherwise is self-defeating—it serves only the interests of those who are motivated to prevent us from understanding the truth of the situation until it is too late to do anything about it.

As members of the public, we are already burdened with far more than our share of disadvantages when it comes to everything involving alien contact. We cannot afford to add self-delusion and wishful thinking to the list of liabilities and obstacles which already stand in our way.

These books of mine tend to contain quite a bit of discussion about the negative aspects of contact between humans and aliens. There is a good reason for that, but I do not want my intentions here to be misunderstood by the reader.

As much as it is possible for me to do so, I do not have an "agenda" when I go about discussing and analyzing the situation. My goal is to be as accurate and thorough as possible, and to avoid filtering my viewpoint through the lens of personal bias one way or another. I do not set out to organize or analyze the information in a way which will end up making it appear that all alien contact is of either a positive or a negative nature.

Clearly, there are many examples of both positive and negative contact. There are many as well where the agenda and intentions of the aliens which are involved in certain incidents are unknown to us.

All of us would certainly hope to find that alien contact is, by and large, a positive experience. I will not, however, assume that this is the case without having good reason to do so. Likewise, I do not presume that aliens are hostile or dangerous without good evidence to back up such a point of view.

As much as it is possible for me to be, I am neutral when it comes to these things. My interest is not to be found in defending certain alien races, or in portraying them negatively to suit my individual opinions in some way. Doing either of those things would result in a biased and ultimately inaccurate view of the situation, and that is something I try very hard to avoid.

Having a personal agenda in terms of wanting to find certain motivations or intentions regarding alien visitors to this world, or wanting to convince others of the same, is counterproductive. Our subjective opinions, hopes or agendas are completely irrelevant when it comes to these things—the only agendas that matter are those of national governments and of the aliens themselves. Whatever we may hope for, or seek to find, makes absolutely no difference—it will not affect the reality of the situation in even the slightest way.

When those hopes are projected into the dialog, the only thing which can result is inaccuracy, confusion and mistakes. To the best of my ability, I hope to provide the reader with an honest, reasonable and unbiased analysis of the topics we examine. I want to understand the truth of things, whatever that truth may turn out to be, and to help my reader do the same.

Though I will never even come close to having all the answers, I do everything in my power to make sure that the questions I *do* find answers to are answered correctly. That is, it seems to me, the best one can hope for.

CHAPTER 9: INTO THE VOLCANO

"The general who is skilled in defense hides in the most secret recesses of the earth."
--Sun Tzu

"Open eyes, but you're sleepin',
You best wake up, before tomorrow comes creepin' in"
--Grand Funk Railroad

In the previous volume, I described a sighting of an alien spacecraft I had which occurred in the year 1984. For literally decades I scoured the records of UFO photographs, trying in vain to find a picture that matched the craft I had seen. Finally, a couple of years ago, author and radio host Debra Jayne East managed to find one, much to my surprise. It was a hieroglyph from an ancient Egyptian temple, which is now kept at the Rockefeller Museum.

It was an exact match. And, on either side of the craft depicted in the hieroglyph, were depictions of what were obviously Grey aliens. After all those years of wondering who it had been that was piloting that craft I saw in 1984, it seemed that I now had my answer.

Not long after that, another piece of information made its way to me from a former member of the black ops community. I was asking him if he could confirm rumors that there is an underground connection which runs from Madigan Army Hospital, just south of Tacoma and part of the Fort Lewis complex, and a subterranean alien base.

I do not know whether such a connection exists, it is simply a rumor which has gone around for years. It is said that the staff tend to wander around, appearing as though they are in a daze…or have been subjected to mind control. It is also said that there are underground levels at Madigan which are impossible to access without making use of a specially-keyed, highly-secured elevator. This is the subject I was enquiring about, simply as a matter of curiosity.

The source had no information about that, one way or another. Somehow, the name of my hometown happened to come up in our conversation. I was raised near the city of Yakima, in Central Washington State, about 140 miles east of Fort Lewis.

All my life, when I was growing up, I had been able to see a large, dormant volcano called Mt. Adams in the distance. Mt. Adams is the second-highest mountain in the state in terms of elevation, after Mt. Rainier. Due to the shape of the mountain, however, which is over 18 miles across at the base, it is the most massive mountain in the entire Cascade Range, which run from Canada down into Northern California and include Mt. Shasta.

Mt. Adams is unlike the other mountains in the Cascade Range in an important way. Unlike it's cousin to the north, Mt. Rainier, which is a popular National Park and is visited by over a million guests each summer, Mt. Adams is for all practical purposes inaccessible. About one-third of the mountain lies within the Yakima Indian Reservation. The Yakima Tribe traditionally considers the mountain to be sacred land. As such, it is off limits to visitors unless they are given a restricted access pass.

I have heard of some small logging operations which took place on the mountain many years ago, and which were associated with the Yakima Tribe, and I know that the mountain has been climbed a couple times, from the opposite side. But even as a long-time resident of the nearest sizeable city to the mountain, to this day I have never met, nor have I even heard of, anyone who has been given permission to go up on Mt. Adams. For all practical purposes, it is an untouched wilderness and a place nobody travels to. The nearest road which runs past it is located almost twenty miles from the base of the mountain itself. To get closer than that, one would have to use the old logging roads.

Within the State of Washington, this lack of public access makes Mt. Adams unique. Even places like Mt. St. Helens, the site of a tremendous volcanic eruption on May 18, 1980, and Mt. Baker, a mountain in the North Cascades near the Canadian border, have public access available and are popular tourist attractions. Mt. Adams, on the other hand, the most massive mountain in the State of Washington, sits isolated and alone, free from the pressures associated with public access. Indeed, it is treated as the sacred mountain the Yakima Indian Tribe believe it to be and is left alone. As it was thousands of years ago, it is today.

I never had any further thoughts about that mountain…until that day I spoke to the black ops veteran and mentioned the city of

Yakima. "I don't know about Madigan," he said slowly and carefully, "but there is an underground alien base near Yakima."

In a matter of about a second, information which had seemed to be unrelated and unconnected suddenly fell into place. Suddenly, I knew. I knew where that base *had to* be…and I knew the type of aliens which inhabited it…and, now, I knew where that craft I saw in 1984 had almost certainly been based out of.

"Greys?" I asked softly.

He nodded. "That's right. Greys."

He did not know the exact location of the alien base itself, only that it was near Yakima. But I suddenly *did* know the location of that base! It had to be located inside that mountain! It was the only place which made sense, the only location in the region where an alien base could exist and regularly have spacecraft enter and exit the base without attracting the notice of the locals.

Indeed, a base inside Mt. Adams could operate without being noticed at all! In fact, it's existence would never even be so much as suspected by anyone in the area, because although the mountain itself is always visible in the distance, nobody gets close enough to it to be able to witness any alien spacecraft which may be entering or leaving an interior base there, especially if the access point was high up on the mountain. Not only that, the fact that the mountain was a dormant—but still active—volcano meant that it would provide all the heat that a high-altitude base would require.

All these thoughts fell instantly into place in my mind, as soon as I heard the words "there is an underground alien base near Yakima". And, knowing as I did then that the craft I had seen so long ago was surely being piloted by the Greys, I could infer who that base belonged to…and where that craft had surely been launched from that evening, and returned to later.

I had no way of knowing precisely where the entrance to that underground alien base might be on the mountain itself…but I was certain it had to be up there *somewhere!*

In fact, the Yakima area does have many sightings each year. It is also the home of what is considered the first "modern" UFO sighting, which occurred in the year 1948 and was reported by a pilot named Kenneth Arnold (this is also where the term "flying saucers" originated).

99

Arnold had taken off in his private plane from the Yakima Municipal Airport that morning. And...wouldn't you know it...had been flying in the direction of Mt. Adams—which is located only about 35 miles away—when he encountered the unidentified aircraft which he reported that day!

The existence of an underground alien base near Yakima is completely unknown to the residents of the nearby cities, and most likely to the UFO community in general. The fact that it was located within Mt. Adams itself is something nobody would have any reason to suspect.

About a year ago, a certain person whom I will not name happened to attend a seminar which took place even closer to the mountain. Mt. Adams, in fact, dominated the skyline from the location of the conference. As he was preparing to get ready to leave, on the final day of the conference, this person decided that he would take some photographs of the mountain, which is spectacularly beautiful and was in such proximity to him at the time.

When he looked at the photographs he had just taken, several minutes later, he got a shock: purely by accident, he had photographed a large entrance portal opening, near the top of the mountain. It was in a location which appeared to be nothing but the rocky face of Mt. Adams, up near the summer snow line. But part of that rocky face had moved. It had opened to reveal what had every appearance of being the entrance to an alien base, located inside this enormous, dormant volcano.

For the record, the photographs that were taken that day were not faked or altered in any way. As soon as he had looked at them and realized what they showed, he immediately posted them online in their original form.

As it happened, I saw them within probably five minutes of the time he posted them, because I was following his reports from the conference he had been posting to his Facebook page. There was no trickery involved, and there was no mistaking what those photographs showed.

The information which had been shared with me by one of my inside contacts had, much to everyone's surprise, been verified by a stroke of luck, aided by the fact that someone had been using a high-quality, professional-grade camera and had taken some photographs with it in the middle of the day in perfect viewing conditions, a sunny cloudless day with nothing between himself and the mountain but a little bit of still air. The photographs taken on that beautiful Sunday afternoon are, in my opinion, the evidence which locks it down and confirms its existence.

Many UFO books consist mostly of a rehashing of long-ago events, the details of which are already well-known to most of the people who will end up reading them. We hear endless retelling of, for instance, the Roswell Event, and patrolman Lonnie Zamorra, and the ill-fated flight of Captain Thomas Mantell and his squadron. We hear about the Rendlesham Forest incident, the supposed messages of the Pleiadeans, and speculation about alien abductions by people who, for the most part, are not abductees themselves and therefore lack the ability to truly shed much light on the topic.

Books such as these are, quite simply, written with the intention of selling a lot of copies on the mass paperback market and earning those who write them a handsome reward in royalty payments from an information-starved public. It is very rare that any of them are responsible for breaking anything which is truly new information, or which adds much of value to the already-existing database of knowledge.

This book, however, and those which make up the rest of the series it is contained in, are of a different type completely. I have no interest in re-telling stories which have already been told a hundred times, or of collecting royalties from books which contain nothing of lasting value to their audience.

As it happens, in this manuscript, the location of an alien base has been revealed, along with the location of its entrance and the identity of the alien race which is associated with it. Combined with my previous report about the 1984 sighting, one can even be informed about the appearance of at least one of the spacecraft types which is flown from that base, in intricate detail because that sighting lasted for over 40 minutes.

Unlike the mass-market pulp books, a valuable addition to the database and a lasting contribution to the field of ufology in general has been made. Regardless of what others may think of this book, that is something which I find to be personally very pleasing, because it is precisely the sort of thing I had hoped for and intended when I undertook the process of writing this series of books about three years ago. They are, it seems, successful in achieving their goal, whether they ultimately become best sellers or sell few copies and are overlooked by most members of the public. It is things like this which serve to keep me motivated to do what I do.

Making things even better, from my point of view at least, is the fact that this is not the only information contained within these pages which will prove to be previously unknown, accurate, surprising and valuable additions to the overall UFO database. There are at least two other instances to be found here to which that description can be applied.

Writing books about alien contact is not a very financially rewarding undertaking. If one wants to get rich as an author, this is not the first choice among topics they should write about—or the second, or the third. It must be done because it is a labor of love and the fulfillment of a responsibility to others. If that is not the case, if monetary reward is the primary goal, a subject with a much broader appeal should be selected rather than this one.

We must, then, take our victories where we find them, and draw our motivation from other things. Chapters such as this one help to motivate me, and to make me feel that this thing I have spent so long creating has been worth the effort, the risk and the frustration which has gone along with it.

I thank you for giving me the opportunity to do this, and the positive feedback which resulted from the first book means more to me than I can put into words. It validates the vision I had in the beginning and demonstrates, I hope, that these books truly are not written with my own interests in mind.

They are written for you, in the way I think you would want them to be written, because you deserve to be able to read something other than mass-market paperbacks. You deserve to get your money's worth, when you purchase a book. I hope you will feel I have given you that, at least.

CHAPTER 10: FOURTH ROCK FROM THE SUN

"It's not going to do any good to land on Mars if we're
stupid."
--Ray Bradbury

"Mars ain't the kind of place to raise your kids,
In fact, it's cold as hell."
--Bernie Taupin

Virtually nothing we are told by NASA is true. It is all carefully manicured disinformation, designed to give the appearance of fact-based reports while making certain that we are prevented from learning the truth about what is taking place within our solar system.

As far as propagating disinformation while maintaining a high-level of public acceptance and trust is concerned, NASA could be described as the most successful long-term hoax in the history of the Space Age. For well over a half-century, it has carried out its mission of mass deception right under our noses. To this very day only a tiny minority of the American public views NASA as anything but a highly-reliable, honorable and well-intentioned organization which has always had our best interests in mind.

If we were to gather all the official reports, mission data and press releases ever released in the long history of NASA into a large pile, set fire to it and forget we had ever heard a single word of it, we would be doing ourselves a considerable favor. We would also be eliminating a major source of confusion and misperception, and something which constantly holds the entire topic hostage to its pernicious secret agenda.

When it comes to the topic of the present chapter—the planet Mars—NASA has managed to construct and maintain what is perhaps its greatest and most successful public deception. It has been so successful, in fact, that virtually nobody in the American civilian population has even the slightest clue what conditions on Mars are really like.

If one were to conduct a public survey, the responses would overwhelmingly indicate that almost the entire nation is firmly convinced that Mars is airless, lifeless, freezing cold and composed of a landscape which is uniformly the color of rust. None of those things are true, yet everyone believes them to be facts which are so well-established as to be beyond any question or doubt by reasonable people.

The sheer quantity of inaccurate beliefs and the depth of confusion NASA has been responsible for foisting on a trusting and gullible public is, indeed, quite impressive. So much so, in fact, that many who read this chapter will find themselves automatically, reflexively disbelieving and rejecting the information about Mars which I am about to present.

Coming to grips with that paradigm shift and incorporating it into one's worldview runs counter to our instincts. We will quite naturally raise all the objections we can come up with, rather than make such a major change to our understanding of something we felt certain we were well-familiar with.

That is something I am powerless to prevent or remedy. All I can do is assure the reader that the level of my research into this topic has been both extensive and carefully-verified. The quality of my inside connections who possess direct personal knowledge and experience regarding the red planet are superb.

These sources are friends of mine and their reliability has been demonstrated over and over through the course of many years. The positions they do or have occupied are such that access to inside information about Mars is not only available to them, it is a regular part of their daily lives and a requirement of their jobs. I trust them completely, and I highly recommend that you do the same—the information they have provided is a result of first-hand experience, and it will not lead you astray.

It is, however, accurate to say that virtually everything you have been taught to believe about the planet Mars is untrue, and the result of intentional deception. Your perception of this close-neighbor is about to be radically transformed in ways you are unlikely to be prepared for.

Such is life. A large part of the value and usefulness of this series of books that I am writing lies in correcting mistakes and educating my brothers and sisters as much as I can. I see it as my responsibility to present the reader with a far more accurate picture of reality than they were given by official sources.

The fact that reality is so very different from what we were taught to believe, while sometimes making the revised picture difficult to understand and accept, is not an indication that it is untrue. It is, rather, a concrete example of the depth and effort which have gone into hiding the truth from us. If our government is willing to take such extreme, pervasive and permanent measures to cover up the situation we face, it goes without saying that the news is—to put it as lightly as possible—unpleasant, unwelcome and not something we would wish for.

One thing about Mars that NASA has somehow neglected to mention to us, for example, is the fact that there are areas of the Martian surface which are covered with easily-recognizable forests. They have been both photographed from close range and personally inspected by military personnel numerous times throughout the years.

Perhaps the reader is already tempted to roll their eyes and disbelieve my words, even here at the beginning of our examination of Mars. If so, they may wish to closely examine the following photographs, all of which originated with either NASA or the American military, and none of which have been officially confirmed or released to the public.

Go ahead. Have a look. Take your time. After all, it is a drastic change from anything you ever envisioned as existing on the Martian surface. These photographs, by the way, were originally all in full-color. I had to convert them to black-and-white images to be able to include them in this manuscript. The name of the person who was originally responsible for finding and releasing them, and his website, are annotated along the edge.

The forests are present in the area near the South Pole of Mars, where water exists for them. The rest of Mars is too dry to support trees or forests.

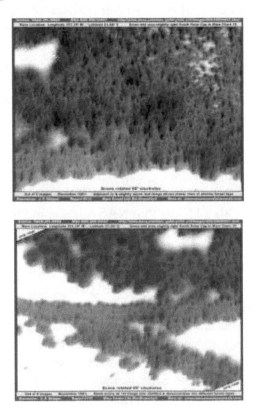

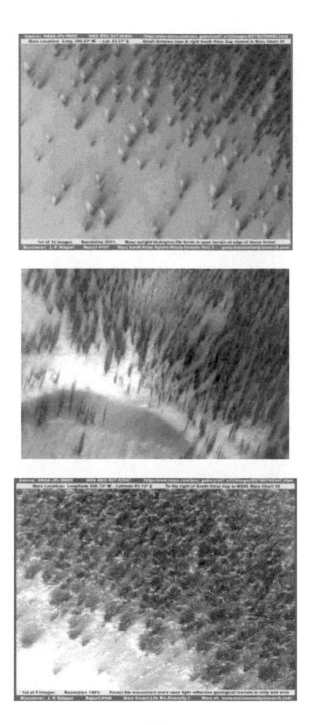

107

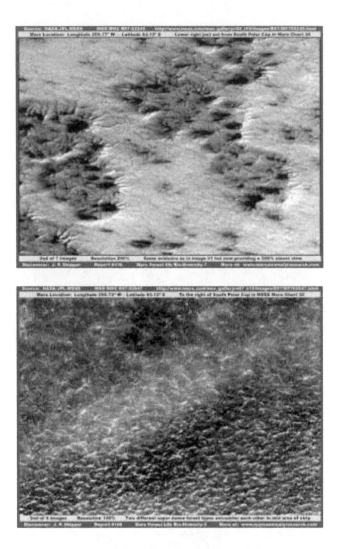

The existence of forests on the surface of Mars directly implies the existence of adequate levels of both oxygen and carbon dioxide in the Martian atmosphere—and, indeed, this is the case. There is a sufficient amount of oxygen in the atmosphere, in fact, that a human can safely breathe the Martian air and survive on the surface for a limited time even without the protection of a space suit.

How do we know this? Because another of the things which NASA and the military neglected to mention to us is the fact that, despite the lack of a publicly-acknowledged manned mission to Mars, we have already secretly been there. We acquired the ability to safely transport both human personnel and a wide variety of materials from Earth to Mars several decades ago. Although the public is not aware of it, elements of our Secret Space Program make the trip on a regular basis.

The base, by the way, is an extensive underground facility. It utilizes high-tech pressure domes equipped with airlocks to allow our personnel to have access to the surface of the world. It has two entrances that I am aware of—a main entrance and a smaller one which is referred to as the "back door".

The first four images which are included below are unreleased photographs of the domes which comprise the main entrance of our secret underground base on the planet Mars.

Although all the photos we are shown of Mars make it appear to have a surface the color of rust, that is just another deceptive tactic used by the government: a red filter is applied to all photos of the Martian surface before they are released. The red filter is also often applied even to photographs which are never released to the public and are classified at a level which ensures they will remain concealed from public view.

A good example of this are the following images, which were given to me over twenty years ago by the gentleman I referred to as "The Colonel" in the previous volume (*"Alien Contact: The Difficult Truth"*). Like the preceding photographs, they were originally in full-color, but were converted into black & white images to make it possible for me to include them here.

These images show what is informally referred to as the "back door" to America's secret underground base on Mars. They were initially processed by having a red filter applied to them. The entire surface of the ground, as well as the domes themselves and even the sky, then appeared to take on a uniform reddish tint like what we see when something has been covered with rust.

I personally know or have known at least three individuals who were stationed at the Mars colony. One of them, who is now deceased, I considered to be one of my closest friends.

I would add that they all agree with the information contained in this chapter. They have verified that all photographs included within it are in fact genuine.

All these people—and, I would presume, many or most of the others stationed there—were assigned to the Mars colony extensive periods of time. Two of them were stationed on the red planet for twenty years. That is far longer than most military assignments, and enough to qualify as a full career.

The third individual spend five years on Mars, and another twenty years serving aboard a large spacecraft which was designed specifically for military use and was in fact a fully-armed warship with interplanetary, interstellar and even inter-galactic capability!

This requires the use of, among other things, a propulsion system which is a direct product of alien technology. Alien technology such as this is incredibly advanced compared to our own. It confers abilities on those who possess it which most members of the public would simply refuse to believe are even possible. Tell the average person off the street, for instance, that we possess spacecraft which can travel to distant galaxies, and then returning home safely, and they would almost certainly laugh at you and roll their eyes.

Coming to grips with the reality of alien technologies which may well be literally millions of years ahead of our own is one of the most difficult and challenging aspects of the study of alien contact. Failure to account for such technology is enough, all on its own, to limit one's ability to understand what is really taking place so severely that it will, in practical terms, be crippled for life.

In exchange for volunteering to spend such a major portion of their lives on a foreign world, far away from their friends and relatives, they were given a special way of making it worthwhile to them. At the end of their twenty-year hitches, they were then transported back to Earth. But that was not all. Using highly classified military technology, were also sent backward in time twenty years and allowed to live the time they had spent on Mars over again, this time on Earth.

The idea that our military can engage in time travel is one which is surely extremely difficult for many people to accept. Even so, I can assure you from first-hand experience (described in the previous volume, in the chapter which deals with abductions) that this is very much the case. I cannot explain how it works, nor can I tell you how they got it or where it came from. I can simply attest to the fact that time travel is indeed possible, that our military and at least some of the aliens can and do make use of it, and that some abductees are involuntarily sent through time for various reasons.

There is much about the topic of alien contact which is difficult to understand, and difficult to accept as being true. That is, however, the nature of the beast, so to speak. If this were not the case, we would most likely not be dealing with highly-advanced alien life-forms in the first place. Just as much of our science would have been difficult or impossible for our ancestors to understand, or to believe is even possible, their technologies are to us.

Though we must of course use great care when deciding what to believe, technologies such as time travel are but one of many things which, though unexpected and bizarre, turn out to be factually true. It is not a matter of imagination or fantasy, it is just a matter of highly-advanced technologies which are based on similarly highly-advanced scientific concepts which we are yet unfamiliar with here in the civilian world.

Our forces are engaged in, among other things, extensive mining operations on and beneath the surface of Mars. They mine for copper, as well as other more valuable metals there.

There, far from the public eye and free from the restraints of any terrestrial laws and customs, these mines—as well as most other strenuous or menial activities which go on there—make constant use of slave labor and have since the colonies were first established. They have no moral compunctions about doing so, and there are no laws there which in any way prevent them from doing so.

This fact provides an explanation for an event described in the previous volume, where I stated that I was transported into the future and used as a slave in a copper mine. Rather than being transported into the future, I now feel certain that I was transported to Mars, and then back to Earth later. The high-tech nature of even the equipment used inside the mines caused me to assume I had been transported into the future. I had really been sent into a facility so advanced that I had no way of knowing it exists at present.

It is probably even now possible for me to name the specific private contractor here on Earth which has been given responsibility for the staffing and operation of these mines. Since that would seem to be a good way to get myself killed for "talking out of class", however, I will refrain from doing so.

Contrary to widely held belief, Mars is not a barren and lifeless world, nor is it simply waiting patiently for humanity to stake its claim of ownership and then begin the process of colonizing and terraforming it. That would, as it turns out, be quite impossible as a practical matter.

Officially, ownership and control of Mars has been established by a treaty which was agreed to by three different alien groups. The planet was divided up between the reptilian race known as the Draco, another—far more primitive—reptilian race known as the Raptors, and a conglomerate of highly intelligent beings which comprise a group commonly referred to as the Insectoids.

All these races will be examined and discussed in detail in the next volume of this series. I had originally intended to include that material in this volume, but in the end the practical limitations of time and space have prevented me from being able to do that. Even with this being the case, I have at least been able to provide the reader with a basic outline of the true situation regarding Mars and our activities there, as well as the identities of the primary alien inhabitants of the planet.

I will also take the time to briefly discuss the entities known collectively as Insectoids. The most well-known member of this collective are beings which are commonly referred to as "Mantids", due to their resemblance to the insect known as praying mantis which exists on Earth.

The Mantids are not identical to our earthly insects of the same name, scaled up in size, nor did the Mantids originate in our solar system. They are highly intelligent and possess an average IQ which the military has estimated to be at least 180.

By comparison, Einstein's IQ was in the low 160's. World Chess Champion Bobby Fischer, who also possessed a photographic memory, had an IQ of 183—about the same as an average Mantid, if the reports about them are correct.

The Mantids are not the only example of highly intelligent alien beings which we classify as belonging to the Insectoid family. It also includes aliens which resemble beetles, and others which resemble grasshoppers.

As with the Mantids, they did not originate here, nor are they exactly like the insects we are familiar with here. Describing them as "grasshoppers" or "beetles" is something which is useful and convenient in many ways but is not technically accurate.

There is a fourth type of insectoid which is known of, too. They are commonly referred to as "spiders" and are said to be incredibly fast. They are distinctly different from earthly spiders, but you will see the resemblance and understand why we call them spiders in a moment. They are reportedly far less intelligent than the other Insectoid forms, and the Mantids train them in a way similar to what we do with dogs here on Earth.

The photos that you see below are, to the best of my knowledge, the first photographs which contain images of legitimate, unaltered alien beings which have ever been published on any forum. Though many photos in various places claim to depict real aliens, virtually none of them really do. They are, rather, a wide variety of cleverly-created fakes which are intended to create and promote confusion and error in the minds of the public.

The photos which follow—one is basically an enlargement of the other—are very different from the other photos which are alleged to contain alien beings. Unlike those images, these are the real thing.

I have spent my entire adult life carefully building and maintaining a reputation for honesty, accuracy, integrity and truthfulness. I will make use of it here and now, by giving the reader my absolute guarantee that I have personally shown the following images to several different individuals who served on Mars during their time in the military. All are well-known to me. They are reliable, honest, intelligent and honorable people whom I am proud and fortunate to be able to call my friends.

They unanimously agreed that the photos show one of the Insectoid forms we refer to as "spiders". All agreed that the photos are accurate, legitimate, and were taken on Mars. I don't know how it can get more real than that, brothers and sisters.

If those folks tell me that we're looking at an Insectoid "spider" form, I am quite willing to put my reputation on the line to fully support their analysis. It is not, after all, something they would get wrong, nor is it something they would have any possible motivation to lie about.

So, then, if you have ever wanted to see what a real, live alien being looks like—and all of us do, that's one thing we have in common with each other—you now can do so for the first time:

117

There is much more to be known about both Mars and various other locations within our solar system which are inhabited by alien beings, and these things will be discussed in their proper time and place. Some of them should not be particularly difficult for even the untrained eye to locate and identify.

One example of such a place might be a moon of Saturn known as Iapetus. It appears to have been welded together. Apparently, there are teams of certified planetary-welding technicians out there that NASA also forgot to tell us about.

CHAPTER 11: COLLATERAL DAMAGE

"Don't confuse something you just pulled out of your ass
with legitimate science."
-- Neil DeGrasse Tyson

"The weekend at the college
Didn't turn out like you planned
The things that pass for knowledge,
I can't understand."
--Steely Dan

Against all odds, there is a large (and rapidly growing) group of people who devoutly believe that the Earth is flat, rather than spherical. Technically, the shape of the Earth can be accurately described as an "oblate spheroid"—a sphere that is slightly flattened on the top and bottom, and slightly wider at the equator due to distortion caused by the centrifugal force of its rotation.

These so-called "flat Earthers" are convinced that the Earth is covered by a solid, impenetrable dome—the "firmament". They do not believe that it rotates. Instead, they have invested their belief in the idea that the sun is far smaller than we have been led to believe. It is so small, according to them, that rather than Earth orbiting around the sun, it is instead the sun which revolves around the Earth each day.

To prove this idea, the flat Earthers state that there is no curvature of the Earth. They claim that the horizon is completely flat, and that this can be confirmed at altitude from an airplane. They are usually quick to mention anyone who will listen to them that Earth's flat horizon is something which has been confirmed by several pilots, who have gone on record saying that this is, indeed, the case.

The flat plane of the Earth, according to this idea, is bounded by Antarctica, which surrounds it on all sides, encircling both the oceans and the continents with an enormous wall of impenetrable ice.

The deep conviction that these people hold regarding the flatness of the Earth, as it usually plays out, is far more than just a simple matter of a disagreement, or a difference of opinion among friends. In their eyes, it is an idea which is beyond any possibility of dispute, and their attitude about it takes on the elements of a religious crusade.

They will not consider—or even seriously listen to—any arguments to the contrary, no matter how reasonable or sensible they may be. They believe all those who stubbornly cling to the outdated, patently false concept of a spherical Earth to be ignorant and blind to reality. As far as they are concerned, it is all very scientific and perfectly logical: the matter has been settled, the proof of it is well-established, and that is the end of the discussion!

Because this belief has become so widespread over the past few years, some who read this manuscript may well be members of the group who proudly identify themselves as "flat Earthers". If you happen to be one of them, I want you to be know that I have taken the time to write this chapter just for you! I hope, therefore, that you will be so kind as to do me the courtesy of reading it slowly and thoroughly, because there are several things I feel you should be made aware of, and which are well-worth giving careful thought to.

I would also like to believe that, by this point in the manuscript, we have established a level of communication between us which is sufficiently well-grounded that you can be comfortable thinking of me as your friend. I hope I have demonstrated by now that I consider all who read this book, though we may not be related by blood, to be my brothers and sisters. I am here on a mission of friendship and genuine goodwill, and it is my intention to promote your best interests. I am certain that you, and all my readers, clearly understand that I would never do or say anything which would intentionally harm or endanger you.

The reason I mention this is that, before we attempt to tackle the issues I am hoping you will give some careful thought to, I would like to ask you to do me a small favor. It's not much—just a simple little exercise, really, and it will provide immediate—and permanent—benefits to each of you. It requires only a few brief moments of your time, can easily be done in the safety of your own homes and will not cost you a penny. In addition, I give you my solemn oath that it will not harm or endanger you in any way.

It will, however, require a certain amount of trust on your part. If you will do me this small favor, carefully following the instructions I am about to provide you with, the changes which result from it will begin to manifest immediately. It will improve your quality of life, enhance your self-image, engender respect from all those you meet.

As if that weren't already enough, it will also be of great assistance when it comes to understanding the nature of the world around you, and it may even serve to make you a bit wiser than you were before. All these wonderful and positive things will become yours, my personal gift to you, friend-to-friend, just for being willing to place your trust in me for a few brief moments! That is a hard deal to beat, my friends, as I am sure you will agree!

Here, then, is what I would like you to do:

Begin by standing in front of a bathroom mirror and looking closely at your reflection. Study it carefully for a few moments, until you are certain you are quite familiar with all its intricacies and could recognize it again in the future.

Next, please look directly into the eyes of your reflection, as though you were going to have a conversation with it. It will also prove to be quite helpful if you give your reflection a look of the utmost seriousness and gravity as you do this, as though you were about to impart some great and profound truth to it.

Now, slowly and clearly, speaking to your reflection in the mirror with all the sincerity you can muster, please speak the following words, without breaking eye contact:

"Some people are born into this world with a high degree of intelligence. I am not one of those people. Others are born with only an average level of intelligence. I am not one of those people, either. What I am is one of the most gullible, scientifically-illiterate people in the entire world. From this point on, I shall leave the thinking to others, who are far better at it than I can ever hope to be!"

Having spoken these words to your reflection with all the seriousness and gravity you are capable of, it is now time to ball up your right fist very tightly. When you have accomplished this, I want you to punch yourself in the face, as hard as you can.

121

Oh, come now, my friends! We both know perfectly well that you can punch harder than *that!* Who do you think you are trying to fool here, anyway?

No, I'm afraid that just won't do at all. Try it again, and this time put some real effort into it—don't cheat yourselves! No pain, no gain! Don't hold back!

Excellent! Nicely done! So far, so good!

Now, gaze once again at your reflection in the mirror. Slowly and clearly, looking directly into your own eyes, say the following words:

"Well! I certainly hope I've learned my lesson! Because if not, Derek is perfectly capable of making me stand here, punching myself in the face, for as long as it takes to knock some sense into me!"

I'm sorry it had to come to this—truly I am. But please know it was an act of love, undertaken only because I care so deeply about your welfare and want the best for you! Believe me, punching yourself in the face hurts me far more than it hurts you!

Well…that's not entirely true. In all honesty, it doesn't hurt *me* a bit. Nor, quite frankly, should it. After all, I am not the one who thinks the Earth is flat, so why should I be penalized? I shouldn't be! You, however, should be—and you were!

Life's lessons come to us from a variety of places and in a variety of ways, often when we expect them the least. That is the first thing I would like you to learn from this exercise, and the first benefit it will bestow upon you.

I recommend that you look at the bright side of the whole dreadful affair: you will be given the opportunity to take several valuable life lessons from this chapter, for the price of only one! You have already learned that life's lessons often come when they are least expected.

You have also learned that you really ought to leave the business of thinking to others, rather than attempting to ad lib it themselves and ending up with a big mess to clean up. In addition, you've learned that you can punch a lot harder than you thought you could, and that there is often a considerable amount of pain involved in being a dumbass.

Over the course of the next several pages, it will be clearly and conclusively demonstrated that the Earth is not, as you naively believe, shaped like a dinner plate. It never has been, nor will it ever be. No amount of misdirected overconfidence—regardless of how enormous and deeply-rooted it may be—will ever have the power to change that.

These things will be added to the already-extensive list of valuable knowledge which has suddenly fallen into your lap, as it were. It has occurred in a very short amount of time, with only a minimum amount of effort required on your part.

All things considered, this is surely quite a bargain, as far as you are concerned! From the looks of things, this is turning out to be your lucky day!

Pretty good friends? Pretty good friends!

How, you might be asking yourself, can I be so certain that the world we live on is not shaped like an oversized dinner plate? Well, one reason is a little thing we like to refer to as "gravity". The effects of gravity upon planetary objects are well-known and thoroughly in agreement with a multitude of reliable tests and observations which have been carried out over a period of several centuries, always with the same results.

According to your view of things, however, the force of gravity appears to be entirely nullified. In fact, it appears to be completely non-existent, and is entirely incompatible with the idea of a flat Earth.

If the Earth were in fact flat, it would have no center of gravity, nor would it exert any gravitational effect upon its inhabitants, the moon or anything else. The satellites we launch would not trace orbital paths around our planet, the moon would go flying off into the distance and eventually disappear.

Whenever we bounced a basketball, it would rebound off the floor and continue moving straight upward until it left the atmosphere. It would then undertake a long, lonely mission into outer space—hoping, perhaps, to be able to drum up a game of one-on-one as it passed through the empty blackness between Mars and Jupiter and entered the asteroid belt.

That's not all, either. Gravity is responsible for a tremendous number of crucial things, both here on Earth and all the way across the unimaginably vast distance which comprises the universe as we know it.

Our way, everything appears to work just fine. Your way, nothing works at all. Do you begin to see the problem now?

I should also probably mention that, as a group, you have been notably silent when it comes to the matter of developing and testing a new, previously unknown, concept of gravity. One in which your version of reality would become possible and conform to an entirely new set of gravitational laws, which you would presumably be kind enough to provide us with.

Since we are already on the subject, it would be remiss of me not to mention that your duty to re-imagine the force of gravity from scratch is not optional. It is required. Until this has been successfully accomplished—as I believe I pointed out previously—absolutely nothing in your version of the universe works at all.

I am tempted to make note of the fact that the universe, as you imagine it, exists only in the chaotic assortment of short-circuits and fused neurons which has been conscripted and forced to stand-in as your fevered brains. I presume, however, that it would prove to be an exercise in futility, and that nothing would be gained by doing so.

Things are never quite that easy, are they?

Well…I guess you wouldn't know about that. So, let's just move on and hope for the best, shall we?

Yes. Yes, we shall!

Before we leave this problematic topic behind forever, though, I would like to make you aware that the concept, and general principles of the force of gravity are something you can easily test for yourself, right there at home. You can perform a simple experiment which will be quite capable of either validating or disproving the statements I have been making about the way it affects objects at a distance.

Here is what you do. First, get a ladder and climb up on the roof of your house—do it carefully, to avoid the chance of injury. Next, slowly walk toward the edge of the roof, until your toes are just slightly behind it.

Now here comes the fun part: take a deep breath, close your eyes tightly…and jump forward as far as you can. Heck, if you want to add some flair and personal style to it, you can even perform a swan dive instead of settling for a boring old ordinary jump!

If you are correct about the Earth being flat, you will continue to hang in mid-air, level with your roof, because we know that gravity cannot exist on a thin, flat plane. If, on the other hand, you find that you crash to the ground in a heap—maybe even headfirst, if you chose to do the swan dive—we must conclude that gravity *does* exist, and therefore the Earth is *not* flat.

I could point out that, if one flies due East, they will eventually encircle the world and return to their starting point—without flying over Antarctica. Naturally, I found it to be quite surprising that, despite that fact that this is something which is extremely difficult to not notice, you had somehow managed to not notice it anyway.

After some further reflection, however, and considering the track record of flat-Earthers regarding pretty much everything else in the world, I realized that I should not have been surprised at all by the situation. There are times in life when one must consider the source of a problem and make whatever type of allowances are necessary and appropriate. *

I could also mention that, by your reckoning, there must be a little, tiny sun orbiting each planet. Since those other suns are nowhere to be seen, it must therefore be the case that they are all completely invisible, both to the naked eye and the most advanced telescopes we possess.

Normally, I would consider this to be an idea which is so incredibly, inexcusably stupid that nobody would ever be foolish enough to believe it. Thanks to people such as yourselves, however, I will instead be forced to describe it as being so incredibly, inexcusably stupid that *almost* nobody would ever be foolish enough to believe it.

If I possessed a somewhat cruel and sadistic nature—and, quite frankly, I wouldn't put it past me--I could even explain to you that the

---

*Flat-Earthers: Do not attempt to wrap your heads around the meaning of the final sentence in that paragraph. It is not necessary and will only result in a migraine headache combined with an overwhelming sense of powerlessness and frustration.

whole "flat Earth" paradigm is something which was invented by the friendly folks at a Certain Intelligence Agency. Its purpose was to string together a series of statements which sort of seemed like they might be true, if you didn't examine them too closely, and use them to manipulate you into wasting your precious time believing in nonsense. In a perfect world, you would even be liable to take it one step further than that: you would commence doing everything in your power to convince all your geographically-challenged friends to join you in proudly announcing to the world that you have proven the Earth is flat. *

I will, however, refrain from doing either of those things. I presume it must be rather inconvenient to try to turn the pages of a book while most of your limbs are in traction, and I don't want to cause you any further trouble.

Don't worry, my dear friends. Just make sure not to short me on the special sauce when I order fries, and I'll handle the thinking for both of us. We each have our roles to play in this world. Some are born to write clever, witty books. Others are born to punch themselves in the face.

Now that we have succeeded in determining which of us is which, you will be given the opportunity to learn several valuable lessons, all of which will certainly prove to be quite handy for you later, simply by reading this one little chapter! Could things possibly *get* any better, from your point of view? I tell you truly, my friends: apart from your sainted mother, there is surely no one in this world who could possibly care more about your well-being or do more to ease the path on your journey through life than I have done! **

Pretty good friends? Pretty good friends indeed!

* It is probably not true that the CIA's code name for the project is "Operation Bonehead", since I just now came up with that designation. I have it on good authority, however, that attempting to find additional information about this by doing an internet search for "Project Valedictorian" is unlikely to be successful.

**(As it turns out, your mother was at my house earlier today, and she mentioned you to me. "Watch out for my stupid kid," she said. "He's got the day off from the drive-in, and he's a troublemaker!")

Unfortunately, it seems that I must now bring this chapter to a close by making note of one additional thing. It is a difficult subject to talk about, and one I had genuinely hoped we could avoid. Sadly, despite my desperate prayers, in the end you have given me no other choice than to bring it up.

Listen, I am not here to lecture you, okay? Nor am I qualified to pass judgment on you, and I will not attempt to do so. I hope you will consider this to be just a bit of helpful advice, given in the spirit of love and compassion. I promise I will never mention it again. Fair enough?

Alright. Here is the thing. I have something very important to tell you. I need to be sure you understand it, so please listen closely. Crack is bad for you. It will ruin your life if you keep smoking it.

I know you think nobody will notice…but we do. We notice.

Stop. Please. Just stop. Don't force us to stage yet another intervention.

That is all.

CHAPTER 12: HYPER-DIMENSIONALITY

"Because they operate beyond the known laws of physics, their abilities are not limited by our common assumptions of what is possible; they are capable of far more and use their abilities to achieve a fine level of control over mankind."
-- "Montalk"

"Might as well jump..."
-- Van Halen

When we speak of alien entities, we are not always necessarily referring to beings which originate and exist primarily within our own dimension, as we would conventionally understand it. It has been established through the testimony of multiple trustworthy sources that there are many non-human, intelligent beings which originate in a parallel, or alternate, dimension and have developed the technological capability to pass through into ours.

It is not known to me precisely how many alien races this applies to, or what percentage of the total that might represent. All I can say about that is that it this group appears to include a significant percentage of the total number of alien races which are known to us, and I have no reason to presume that it could not in fact include *most* of them.

I have seen reports which indicate that the race I refer to as Type I Greys—the aliens which were involved in the Roswell Incident in 1947—possess the ability to transmute their bodies into pure light waves (or, according to others, sound waves). By making use of this ability, it has been reported, they have been able to "vanish", apparently into thin air, when confined in holding cells by the military, thus effecting their escape.

I have no way of verifying whether this reported ability does in fact exist as described, or whether it is just one more component of the general disinformation program which was intended to confuse us and waste our time chasing down a dead-end street. The phenomenon has been reported by a sufficient number of sources which appear to be reliable, however, that I think we are wise for the moment at least to give it careful consideration as a possibility, if not the provisional benefit of the doubt.

One of the people who reported this was Milton William "Bill" Cooper, who was one of the most well-informed and influential whistle-blowers of all time. Cooper had a no-nonsense style and a habit of rattling off bursts of information the way a submachine gun rattles off bullets. He had so much data at his fingertips, and so many distinct aspects of both the alien contact phenomenon and the subject of conspiracies in general that he could address, that members of the audience had to pay close attention as he spoke, lest they let their attention wander for a few seconds and, as a result, completely miss a key point he had made during that time.

Bill Cooper was armed to the teeth with facts and inside information--he had no time for nonsense and no patience with those who engaged in fraudulent testimony. In my opinion, he stands as one of the most reliably accurate and scrupulously honest sources of inside information in the history of ufology. Countless times he was responsible for coming forward with information which was later— sometimes much later—proven to be accurate.

If William Cooper said something is true, but it is something which I can neither confirm, logically explain or even claim to be able to truly understand, I tend to give Cooper's information the benefit of the doubt. I consider that my inability to explain or understand it is *my* problem, not a problem with the information itself. One does not contradict a man like Cooper lightly. *

Let us then, for the moment, stipulate that this reported change of form, from a physical body like our own into a pure wave-form state, is not impossible and may very well take place just as described. Is this what is meant by hyper-dimensionality, when the term comes up regarding alien beings?

No, it is not. It appears to be something of a completely different nature, as best I can ascertain, and—as with the ability to "jump through" into other dimensions, it is highly likely to be something which is accomplished using advanced technology rather than being

---

*Near the end of his life, however, Cooper did make one serious misjudgment. He came to believe that aliens did not really exist, and that it was all an extensive deception being carried out by the government. This is not actually the case at all.

a natural ability. For such a thing to be a naturally-evolved characteristic of an organism of any kind is a difficult idea for me to imagine, and an impossible one for me to accept without solid evidence to demonstrate it.

I am quite capable of thinking outside the box. I am also aware that there is much about both aliens and their technical capabilities which seems to be nothing less than bizarre to us. But that idea is, for me, a bridge too far. I won't go there without first being shown a much better reason to do so than I have seen at present.

Whether true or not, it is not relevant to our discussion of hyper-dimensionality as it is normally defined. The kind which is normally referred to is a parallel universe, with all the same planets and stars we are familiar with, but existing at a vibrational level which makes it invisible to us and allows it to occupy the same space we occupy, at the same time we occupy it. The easiest way to visualize it, according to my understanding of things, is to imagine a stack of blankets, one on top of another.

It is believed that the American military, in cooperation with NASA, has developed the ability to "jump" across dimensions themselves, at least to a limited extent. My information about this is several years out of date, but the last I heard they had only managed to stay in a parallel dimension for a matter of a few seconds after they popped through into it, then were involuntarily snapped back into our own dimension.

It is important to note that many alien civilizations are believed to be both hyper-dimensional and interstellar, meaning they originated in a parallel dimension, on a planet which orbits another star. It is even believed—and I have no reason to think it's not true—that some are hyper-dimensional, interstellar time travelers.

Just when you thought things couldn't get any more complicated than they already were, right? I know. I hear you. Nobody said this was going to be easy—and it isn't!

The specifics of hyper-dimensional existence, the details of inter-dimensional travel and the effects it may have on those who undertake the journey are speculative, unclear and often contradictory, depending upon which source one happens to be listening to at the time. They cannot *all* be correct…and it's possible that *none* of them are correct. I would like nothing better than to fill you in on the inside scoop regarding all these things, but I am unable to do so (at least at present). It is a tricky and treacherous trail we walk when we attempt to take on this aspect of alien contact, and I would rather say too little about it than say more than I should, only to find out later that I was mistaken and have unintentionally misled those who were good enough to purchase my book and read it.

Hyper-dimensionality will surface again, in the next chapter, and some additional details and beliefs about it will be examined there. For now, I think the purpose of this chapter--which was to make those who were previously unaware of the hyper-dimensional nature of certain alien races at least somewhat familiar with the concept—has been adequately achieved. At the very least, it has been discussed to the limits of my admittedly-insufficient level of knowledge.

I hope to continue to learn about this fascinating—and inherently dangerous—aspect of alien contact and return to it later, in some future book. For now, though, the reader knows as much as I do (and some of you who are reading this probably know quite a bit more than I do) about the general concepts of hyper-dimensionality. I am left with little choice, it seems, but to move on to the next chapter and the next topic.

CHAPTER 13: SHADOW BEINGS & HAT MEN

"The Hat Man almost exclusively wears a fedora and is much more unnerving. People report stark terror when confronted by this entity and feel it is feeding off their fear…this type of Hat Man is not a ghost; it is something much more malevolent."
--Stephen Wagner

"Can you feel me?
I'm danger, I'm the stranger
And I, I'm darkness
I'm anger, I'm pain"
--Ronnie James Dio

An enigmatic and poorly-understood race of what are apparently alien beings are commonly called shadow beings or hat men. Some do not believe these entities truly exist at all. I am here to tell you that they are quite real and that I have personally seen and had at least one interaction with one of these beings.

I would love to be able to inform you that they are friendly in their relationship with humanity, rather than add yet another race to the ever-expanding list of beings who appear to be hostile and intrusive. Sadly, I am unable to do that, and the hat men must therefore be added to the list of hostiles. I can also say with absolute certainty that my brief encounter with one of these beings is the most terrifying experience of my life and something I would not wish on my worst enemy.

These beings are often seen taking the shape of a man whose appearance is a featureless deep black color and who appears to be wearing a long trench coat and a fedora. None of the features of either the being itself or the clothing it appears to be wearing are visible at all—they appear to us only in the form of what appears to be a two-dimensional silhouette. Why they would be wearing a two-dimensional version of a trench coat and a fedora is completely beyond me…but the one I saw was, indeed, wearing a hat of this description and fit the classical description of a "Hat Man" in every way. I would compare them to the characters in an old cartoon strip which was called "Spy vs. Spy".

The two-dimensional appearance that is often reported after sighting one of these beings is perhaps the most puzzling and unaccountable aspect of what is apparently their natural form. We do not know of any possible combination of natural evolution or physical attributes which could either produce such a being or allow it not only to exist, but to display an apparently elevated level of intelligence and an unmistakable sense of purpose.

I do not understand how something can be alive and intelligent, yet at the same time exist in only two dimensions. To us, this is something which does not appear to make sense. There are numerous issues associated with the idea of living beings which do not have at least a three-dimensional form, and some of them seem to be problematic enough to make one consider the possibility of discarding the whole idea out of hand.

We are unable to legitimately do so at present, however, because these beings which have a two-dimensional appearance have been reported by far too many people, in far too consistent a manner, to simply dismiss them without a compelling reason to do so. So, for now, we are left with the unenviable task of trying to either explain or categorize them to the best of our ability in a way which makes sense.

Could they be a projection of some sort? Perhaps something like an astral body, sent out to perform certain functions or missions, but limited to only two dimensions? Could the reports that they exist in only two dimensions be in error? Could they exist in three dimensions, but take on the appearance of being completely flat? Do they exist only on the ethereal, or energy, planes and not in the physical world?

This is a likeness of an entity known as "the Hat Man". The one and only time I had an encounter with this type of being, its appearance was exactly as shown in the drawing above: totally black, wearing what appeared to be a fedora-type hat and a long trench coat, and I was unable to determine whether it was manifested in the dream state in three dimensions or only two.

This entity was, by a large margin, the most terrifying thing I have ever encountered. It radiated a powerful sense of malice, pure evil, hatred, and relentless, pitiless, distinctly malevolent intent. It made me feel, quite literally, as though I was in the presence of the Devil Himself.   Those are questions I am unable to answer with certainty at present. I tend to doubt that the reports are mistaken concerning the "Hat Men", because the one I observed appeared to be two-dimensional in nature, and this is something which is difficult to mimic or counterfeit. Something either has height, or it doesn't--there is no in between, and no grey areas are involved.

I do not believe that the "Hat Men" are the same as what are commonly referred to as "Shadow Beings". That is my opinion, and I state it as nothing more than that. I have seen several purported examples of these shadow beings which appear to have been captured by video cameras. I am very far from being an expert on either video cameras or the numerous ways video clips can be altered, modified or completely forged to produce the results one wishes to produce. Having said that, it seemed to me that the examples I saw appeared to be legitimate and were very strange indeed.

The beings in these clips did not look like the "Hat Men" at all, though they were completely dark and featureless. They were not wearing the distinctive fedora and long jacket that the Hat Men are known for. In fact, I couldn't say for sure that they were wearing anything at all, because they were pitch black—like shadows in motion—and no details were visible on their bodies that might have indicated clothing.

They were upright, like we are, with two legs, two arms and a head. They appeared to be running, moving extremely quickly, passing through the field of the video in only a second or two. One clip I recall had one of these beings appear to be running along an aisle in a crowded stadium at a soccer game, making no contact with the people he ran in front of as his body appeared to sprint in front of a section of seats. Another seemed to be running along a small trail in a jungle, beside a stream, and was captured by some young men. It was perhaps in Brazil, or another Central/South American location. One of the boys was walking down the trail and the being came quite close to him, frightening both he and his companions with its sudden appearance and disappearance.

There is another aspect to some of these video clips which is both mystifying and highly alarming. Though there are many instances of what appear to be two-dimensional shadow beings which are both reported and apparently captured on film, there are others as well, in which the beings do *not* appear to be two-dimensional.

There are multiple clips in which people are physically assaulted by what can only be thought to be shadow beings. In some instances, the people were walking alone down a hallway, and one of these things seemed to materialize out of thin air from a shadowy spot on the floor or a wall. They then attacked the human, sometimes literally knocking them to the floor on their back and beating on them with their shadow fists.

There was no apparent reason for these attacks. What was the shadow being planning to do, beat the guy up and then steal his wallet? Use the stolen credit cards at some shadow ATM in another dimension that accepts American Express? Anally rape them with a shadowy penis? Help them get to their feet and then say, "welcome to the NFL!"? Tie them up and then torture them by poking ectoplasmic needles into their eyeballs until they reveal the secret of who shot J.R. Ewing? What possible point could there be to these kinds of attacks?

While we are on the subject, why does it seem that everything which comes here from another dimension has such an unquenchable hatred of humans? Are we truly that despicable as a species?

Is there some overwhelming, inescapable hatred and thirst for vengeance that would cause something to develop the capability of constructing a dimensional portal, then step through it and hide in the shadows, awaiting its opportunity to assault a total stranger? If that is what is taking place, then we are dealing with something which is far beyond my ability to make rational sense of, and the only reasonable course of action which remains is to wave the white flag of surrender and hope that my death will be quick and painless.

I mean, seriously: who could really hate anyone that much? Try to imagine a shadowy version of Robert Oppenheimer ordering the staff of Los Alamos to direct their efforts to the task of developing portal technology with all possible speed "so I can step into another dimension and deliver my final vengeance upon Kirby, the hall-walker!" Then try to imagine the staff of Los Alamos agreeing to do so.

That is the kind of situation that would had to have taken place in some other dimension. And are we then to imagine that beings which are mentally unhinged enough to want to take inter-dimensional revenge on some random guy walking down a hallway, are intelligent enough to develop portal technology in the first place? The human mind is simply not equipped to deal with such possibilities.

Both the shadow beings and the so-called "Hat Men" represent entities which I do not understand well enough at present to be able to describe them further. Neither can I offer any useful hypotheses about them, other than the thoughts I have already presented in this chapter.

They do not appear to have much of anything in common with any other types of aliens I am familiar with. It is possible that they could represent something quite distinct from them in terms of both their origin and intentions. If that is so, I cannot say what that origin may be, unless they both represent either psychic projections of prodigious power, or the shadows of beings from some other dimension of reality, as perceived in our own.

As best I can determine at least one—and possibly both—of these entities are certainly malevolent in the extreme. I can absolutely promise you that a close encounter with the Hat Man—even one which takes place in the dreamscape, rather than the physical realm—will prove far more terrifying than any other contact experience you will ever have.

Wherever it may be from, and whatever it may ultimately turn out to be, the Hat Man is without any question highly dangerous to those who encounter it! I believe will most certainly inflict maximum harm upon those it encounters, if it is able to!

With that in mind, the best advice I can give right now to anyone who might be unlucky enough to encounter either of these beings is simply this: ESCAPE! Get as far away from them as you can and do it *fast!* Do not hesitate, nor attempt to interact with them in any way whatsoever! Just break contact and put as much distance between them and yourself as possible—IMMEDIATELY!

I hope to have the opportunity to learn more about these mysterious—and highly unconventional—beings in the relatively near future. If I do, I will certainly pass that knowledge along as quickly as possible. Either way, I sincerely hope none of my readers ever have the misfortune of encountering one themselves.

CHAPTER 14: THE ENGINEERS STATEMENT

"Scientists investigate that which already is; Engineers create that which has never been."
--Albert Einstein

"There's no way out of here,
When you come in, you're in for good.
There was no promise made,
The price you paid, the chance you took."
--David Gilmour

The person whom I will refer to here as "The Engineer" could be either a single individual, or a conglomeration of several people who are known to me. For the sake of their safety and privacy, I will not reveal which of those things is the case. I will say only that this person—or persons—were affiliated either with NASA, the military, or governmental intelligence agencies, that they were highly experienced in matters dealing with black operations and had an average experience level of somewhere around twenty years of service at the time I spoke with them.

Because I am unable to identify them personally, the reader is free to choose whether to accept the material in this chapter as being authentic, or to discard it, either in part or in full, as seems best to them.

Being unable to identify them by name is unfortunate, but unavoidable. Because I wanted to include this material in the book, I have chosen to do so in this fashion, despite the disadvantages which are inherent when one uses anonymous sources. Though the sources are—and must remain—anonymous as far as the public is concerned, they are not anonymous to me. They are people whom I know well and have known for many years. They have proven their trustworthiness on many occasions.

I do not automatically presume that everything they might say is completely accurate, and it is known that virtually nobody is given complete access to the unredacted data. The reader would be well-served by keeping that caveat in mind as they read through the material in this chapter.

To the best of my knowledge, the information was given to me in good faith by people I have had a long relationship with, and I believe it is as accurate as it is within their power to make it. If I felt otherwise, I would not permit it to appear in a manuscript which has my byline attached to it. We do the best we can, and sometimes the best we can do turns out to be something like this.

Even so, I feel confident that the reader will not be led astray if they choose to accept this information at face value. It is not particularly Earth-shaking, though it has never been published before, and I think that this fact lends additional credence to the idea that it is not disinformation, nor is it intended to confuse anyone.

The following transcripts do not represent a single conversation but are rather a conglomeration of numerous private conversations which took place via the telephone, e-mail and personal, face-to-face contact between me and the person or persons involved. They can, in my opinion, serve as an example of the fact that one can never be completely certain about who knows what, or where more pieces of information may be found.

The Engineer: "One of the technologies which were gifted to us by the visitors deals with the creation of a cloaking device for our aircraft. It can literally bend light *around* the body of the craft in question. When this system is operational, the result is that the aircraft will disappear from visual sight. It could be right over your head, and you can look right at it…but all you will see is an apparently empty sky. Your eyes will take in the color and appearance of the sky which is behind the aircraft. If the sky is clear, you'll see a blue sky. If it's a cloudy day, you'll see a grey sky, or fluffy white clouds…but no aircraft.

"This is one reason that the stealth planes and other, even more clandestine, platforms cost so much money to produce. It is not the only reason, because sometimes other "technologies of unknown origin" also come into play. I don't know for certain what or how many may be in place in any single design, but they could include things like propulsion systems, instrumentation, defensive

140

capabilities, weapons systems and other things which are of a distinctly unconventional and exotic nature.

"I remember that you once expressed sadness that the SR-71 Blackbird had been retired from service. I sympathize with that feeling, and I share your opinion that it was surely the most beautiful and artistically well-conceived aircraft ever built.

"One may reasonably wonder why it was withdrawn from service, prior to having a more capable aircraft ready to replace it when it was no longer available. The reason for that is that there *were* more capable aircraft, already built and tested, ready to replace it immediately upon its retirement. As is often the case, the public was unaware of those aircraft designs, and many people wrongly assumed that we had retired the Blackbirds without being able to replace them.

"As anyone involved with the military would surely tell you, that is simply not something which would be allowed to happen. The SR-71 had capabilities which were far beyond anything available to its contemporaries…but the designs which replaced it are a quantum leap forward in technology. They are so advanced, when compared to the Blackbird, that they caused the Blackbird to become obsolete in comparison, almost overnight.

"There was no reason to continue with the Blackbird program, and every reason to spend the time and money instead on performing additional research and upgrading the newer vehicles as necessary.

"I am not at liberty to speak about the entire array of the technologies and capabilities possessed by the current generation of stealth aircraft, nor could I produce a complete list of them in any case. But I can tell you that some of them are well beyond anything you might expect. They represent a differential of probably thousands of years, in a single generation of clandestine development.

"Derek…some of the things these new planes can do are almost beyond belief, from the point of view of conventional science. That is one of their great advantages: they have capabilities which nobody will expect, and can therefore carry out mission assignments which, likewise, nobody will expect to be carried out. They cost a bundle of money to build, but they give us certain advantages which it is felt more than justify the cost.

"The strength of our conventional military forces alone serves as a deterrent to other nations. Nobody in the world is going to

141

intentionally start a war against the United States, because our ability to project power and to bring state of the art technology to the battlefield is, at the current time, overwhelming no matter which foreign nation might be under discussion.

"It won't be that way forever, of course…but it *will* be that way well into the foreseeable future, and concerns about nuclear warfare between us and Russia, or us and China, are vastly overblown. Russia and China are many things, but they are not suicidal. They may attempt to subvert us, or utilize financial pressures to weaken us, but a direct military confrontation is out of the question for the time being."

"We have an underwater base in Puerto Rico, located beneath a large, swampy lagoon. This is a human-controlled base.

"The local area, on the surface, is home to many UFO sighting reports. The people who live nearby often report seeing disks and other strange-looking craft rise out of the lagoon and then fly away at high speed, or vice-versa: sometimes they approach the lagoon from the air, then pass beneath its surface and disappear. To the best of my knowledge, all these craft belong to us.

"They are probably utilizing technology which was either reverse-engineered from alien tech or was loaned to us by the visitors. So, though they appear to be alien spacecraft to the casual witness, they are flown by military pilots and owned (or borrowed by) American black ops forces."

"I see. Very interesting indeed!"

"Once, though, I saw a hybrid there, in a lab."

"Ahh, you were in the same room with it?"

"Yes."

"Can you tell me about it?"

"I was standing back about four feet from it. It was about 5'6". Weighed around 130 pounds, from the looks of it. The skin had gone grey from death, but there were only four digits on each hand. The head was enlarged but had hair. The eyes were bigger than ours, but not as big as is reported on Greys, and they were more rounded.

142

"The toes were spatulated, flattened out. Spine showed through skin as if it were meant to, not because of being skinny or dead. The ears were just little nubs. Nose was turned up, with very thin slits for nostrils. It was found wrapped in fibrous aluminum in 300 feet of water."

"It was dead when it was found?"

"Yes. It was wrapped just like a mummy, including arms folded over the chest, but the cloth was made of aluminum fibers or something similar. It was found in 300 feet of water, just outside the canal, which leads to the open sea.

"We were all shown it because some idiot had gotten snapshots of it. They arrested him, but they wanted all of us to see it first-hand, so in case his pics got out we would know the truth. It was near our underwater base in Puerto Rico. Guess what they're calling that aluminum cloth?"

"I have no idea. What?"

"The Shroud."

"Seriously?"

"Yes. The Shroud."

"Well…it sounds like a perfect match to the descriptions I have heard from other sources regarding hybrids, no question about it."

"I have never heard a description of a hybrid. I just said what I saw."

"Well, that is it, what you just said. Some are said to look more like humans, others look like what you saw."

"I'm curious about the spatulated toes. What would cause that, other than very high gravity?"

"I don't know, but it can't have been caused by high gravity. The Greys are not built for high grav."

"Right. So, what would cause spatulated toes?"

"Hmmm…"

"It was not a Grey, though. Spatulated toes imply lots of weight for long periods of time."

"It's not a Grey, but a Grey is what was used to create the hybrid, and humans do not have them either."

"I haven't a clue."

"I have never heard of spatulated toes before."

"Yeah, they were flattened out and spread."

"That is very odd, okay. Don't know what could cause that, really. Could it be a genetic defect, maybe?"

"Camels have the same kind of spatulated feet. These were too uniform. I don't think it was a defect."

"Could it be from feet that are partly from Greys being forced to support too much weight? A hundred and thirty-five pounds is far too much for the Grey anatomy to be able to support. It's like a full hundred pounds more than their skeletons are designed to handle."

"That would make them look flat from abuse. These looked like they were meant to be that way, because the nails grew from side to side too. If it was caused by abuse, the nails wouldn't spread also."

"True. They were not webbed at all, were they?"

"No. No webbing."

"Five toes? Four fingers?"

"Four fingers, five toes—but no big toe. All the toes were identical in length and thickness."

"And there was no thumb on the hands, right?"

"Right. No thumbs."

"There is an underwater base on the sea floor, off the coast of J-burg, South Africa. I have been there twice. The last time I went down, they were in a hurry. They wanted me to get there with all possible speed, so they arranged to have me flown down in the second seat of an F-15, with mid-air refueling taking place along the way. It was the only time I've ever flown in a fighter.

"The pilot made sure I was strapped in tight and told me that he had been ordered to get me down there as quickly as possible. He said that this meant we were going to be flying at well over Mach Two and that there was a good chance that it would make me feel nauseous, that I might even feel the need to throw up at some point along the way.

"It wasn't really that bad, as it turned out. The g-forces were not particularly comfortable for me, since I am not used to flying in high-performance aircraft like these, but I didn't get sick or anything.

"Do you remember when President Clinton was down in South Africa recently, supposedly on official business? Well, the real purpose of that trip was that he was taken out to that base and given a tour of it. The official reason for the trip, the political reason, was just a diversion, an excuse to get him into the area so he could inspect that base.

"As with the base in Puerto Rico, this is a human-occupied and human-controlled base. Some bases are not. Some are jointly occupied by both humans and aliens, as I understand it.

"I have never been to one of those. Not many people have. I suspect they erase the memories of most who have, so they will never remember being there. That's just a guess on my part. But it makes sense that they would.

"You should also be made aware that chemically-based memory erasure is being carried out daily, both to military abductees—who include both members of the civilian population, and members of the military itself, along with its corporate partners. When I say "chemically-based memory erasure", just to avoid any possible confusion, what I mean is "drugs". Specially designed and produced drugs which are by this point in time so highly refined that they can literally determine the number of hours of memory they want someone to have erased very precisely, by altering the amount of chemicals which are introduced into the body.

"If they want you to forget everything that has happened to you during the past eight hours, for example, they know just what combination of drugs to administer, and how much of each will be required to remove that much memory from a person's mind. I do not know the name of the specific drugs involved in the process, only that they are and extremely effective."

"Do the aliens utilize the same drugs, in combination with other things, to erase the memories of those they abduct?" I asked.

"I have no direct knowledge about that one way or another. It would make sense that they probably would—and, for that matter, it's far more likely that *we* got *our* drugs from *them*, than vice versa. It is my opinion that they most likely also utilize hypnosis, post-hypnotic suggestion, neuro-linguistic programming, psychic blocks and triggers, and probably other techniques as well, along with drugs, to delete the memories of abductees.

"But I cannot be certain about any of those things. I am not involved in the medical or pharmaceutical aspects of any of this. I am just making an educated guess, based on various things I have seen and heard from others throughout the years.

"As with everything else when it comes to matters of exotic technologies or non-human entities, information is held very tightly and is dispensed strictly on a need-to-know basis. I had no need to know about any of those things, so I was not informed about them…and of course I knew better than to ask.

"You learn very quickly not to ask questions about this stuff, unless they are questions which must be answered for you to do your job. If you can do your job without being given additional information, that is precisely what will happen.

"Asking questions will not result in you getting any more information, it will only result in your superiors glaring at you and ordering you to drop it, to never bring it up again if you know what's good for you. When your employer's primary responsibility is making sure they always can kill whomever they want to kill, there is no place to go after that other than back to work, and nothing to do other than to never bring it up again.

"Those are the standard restrictions that have always been applied to me. I would be surprised, for that matter, if they didn't also apply to everyone I have worked with along the way, including my handlers and supervisors. It's just the way things are done when you work in black ops, and of course even more so when your function involves interfacing with technologies of unknown, non-human origin."

"I once had the opportunity to board an alien spacecraft during my official duties. The craft was disk-shaped, with an outside diameter of 45 feet. Inside, the diameter was 35 feet. It had a smooth, dark-colored floor. The interior was bare, except for three chairs spaced equidistant in a circle.

"The control panels looked strikingly like the control panels in *Star Trek*. They were low and smooth. There were no buttons, switches or anything else. They were all flat panels.

"There were no colored likes like there are in *Star Trek*. Instead, there were depressions for 'hands' to fit into. Theory is that physical contact with the panels create a type of biological circuit between the panel and the operators.

"The control deck took up half the ship. The lower half of it housed the engines."

"Were they human-sized chairs, or too small for humans?"

"The chairs were right for someone about five feet tall. Six feet would be a real tight fit. From outside, there were no windows at all. From the inside, though, it was ALL windows!"

"Do you mean that you could see through the hull of the craft, in all directions?"

"From immediately above the control panels, which were roughly shoulder level, to a point roughly five feet up the inclined wall, and all the way around the cabin in all directions, it was completely transparent."

"Do you happen to know who designed and built that craft?"

"No, but it was clearly intended to be used by small persons. The height of the ceiling was only about six feet."

"Can you describe the vehicles entrance or hatchway?"

"It had only one entrance. It was a ramp, with a split gull-wing upper half. Imagine two gull-wing doors mounted side-by-side at opposing angles. They reach halfway down the entrance opening. The lower half is a gate that drops down and then extends, forming a ramp.

"The entrance has an internal chamber, like an air lock, which opens on the main deck. I always saw it fully opened up, however I did see it twice when it was being opened and closed."

"Did you get a chance to look at the underside of it?"

"The underside was curved, rounded and smooth. It had no legs or visible supports. Trying to look all the way through to the other side of the craft by looking across its underside was hard on the eyes. It was like trying to look through those wavy heat lines you get on hot pavement. After a minute or two, you have to look away and re-focus your eyes."

"What was the interior light source?"

"There was just light...a pale, washed-out blue...but not dim. It was bright enough to see everything with NO shadows. The light source...I don't know. There were no visible bulbs. The light just seemed to....*be*."

"What did the exterior of the craft look like?"

147

"It was cobalt blue, and absolutely identical—inside and out—to the spacecraft from the movie *Independence Day*. Or, more correctly, the spacecraft in the movie was identical to the real thing. What a great way to hide something. You hide it in plain sight, and then later you can just tell people it's nothing but an old prop they used in the movie."

"Could you tell what the hull was made out of?"

"No. Nobody could, at that time. It had the feel of cold metal but was somewhat resilient.

"There is an optical illusion, as well: it LOOKS like you could put your hand through it, but it's actually solid. It was so smooth, though, that it made its surface appear to be farther away than it really was."

"Did it utilize anti-gravity engines for the propulsion system?"

"Nobody knows, so far. Best guess is pulse-charged magnetics. However, that was a year ago. They may have better answers by now."

"What prevents these craft from causing a sonic boom when they travel through our atmosphere at high speed?"

"Good question! That issue was addressed by the engineering team. Sonic booms only occur when a surface breaks the skin tension of the surrounding air. By using magnetics, the atmosphere is shunted AROUND the ship, so no skin tension is broken.

"This also contributes to higher stability, better navigation and would allow them to navigate through a heavy atmosphere or underwater with no reduction in either speed or maneuverability. It is what we think of as the 'airfoil effect', applied to a spacecraft."

"I was told that we can use their technology to make humans—specifically, members of our special forces—appear to be visually identical to the Greys. That a person can look right at the black ops personnel and they will not be able to tell that they aren't actually looking at an alien Grey. Is that true?"

"I don't know for sure, but I would guess that we can do that, yes. It's believed that the military has employed people of very short stature, for one thing, and have had them dress in the Greys suits as well as wear latex full-head masks that make them look identical to the aliens themselves. We know that their space suits put out a huge electro-magnetic flux somehow. I would think there are probably additional methods available to us that would allow us to disguise our operatives as Grey aliens, too, though I am not familiar with the technical details of how it's accomplished."

"Do you know whether we are cooperating with two groups of aliens who are opposed to each other, and perhaps are even at war with each other?"

"I know we are definitely cooperating with *someone*. Whether they get along with others, I have no way of knowing. I do know, however, that there ARE groups of aliens that do NOT get along...but I don't know whether we are working with one of those groups too or not."

"If I guarantee to protect your identity from disclosure, and invent a fictional name to use instead of your real one, may I include the things you have said here today in a published manuscript at some point?"

"Yes. Absolutely. I know you well enough to have complete faith in you in the matter of protecting my identity and my personal safety.

"I should tell you that you have just been given a great compliment, by the way, because there is no business in the world which even remotely approaches this one in terms of being inherently, unrelentingly paranoid and treacherous.

"My peers and I have long made a habit of refusing to trust those who are not part of our immediate working group with any inside information whatsoever. It is all but impossible to find the kind of people who can be relied on to keep their word to us and protect our anonymity. It is even rarer to find someone who can not only be trusted to keep their promises, but who also has the ability to get the message out in credible manner."

"Thank you, that's a very nice thing to say and I appreciate it more than I can tell you. I have no way to properly thank you for helping to educate me about these things."

"Just get the information out, as we discussed. That will be all the thanks we need, Derek."

"Okay. I'll see to it."

"I know you will, and so does the Colonel. He watched you carefully, from the shadows, until he had seen enough to be convinced. Nobody else was ever able to convince him to trust them."

"Lucky me..."

CHAPTER 15: THE TARD PARTICLE

"The God particle could result in the destruction of the universe. If it becomes unstable, the universe could undergo catastrophic vacuum decay. This could expand at the speed of light, it could happen at any time, and we couldn't see it coming."
-- Professor Stephen Hawking

"Hey you,
Out there beyond the wall,
Breaking bottles in the hall,
Can you hear me?"
-- Pink Floyd

There is not sufficient room in this manuscript to allow me to discuss the CERN project in the amount of detail it deserves to be given. That will have to wait until a future volume. You may be certain, though, that I will eventually get to it—its time is surely coming!

For the moment, I simply want to make note of something the short-sighted, over-confident, world-destroying science nerds who run that show appear to have overlooked. As you are probably aware, they have long claimed to be searching for something they refer to as "the God particle". Now, most professional scientists that I am personally acquainted with are atheists. Shouldn't they, then, choose to refer to the thing they claim to be searching for as "the God does not exist particle"?

I realize, of course, that "the God does not exist particle" is a clumsy, inconvenient thing for them to have to say in every press statement they release. I humbly suggest, therefore, that it instead be named something which is short, concise and—it seems to me—quite fitting, when one considers the apparent possibility that they could end up destroying the universe. I suggest that it be re-designated, using the name I have helpfully contributed, as a public service and completely free of charge, in the title to this chapter.

I fully expect that—as is usually the case—the scientific establishment will spurn my ever-so-appropriate suggestion, and that it will be ignored in favor of the name they have already chosen, despite the apparent inability of their chosen moniker to even co-exist with their stilted and myopic view of our universe.

I also expect that—as is also usually the case—there will come a day when they will wish they had listened to my good advice. I think they will someday regret the fact that they stubbornly insisted on calling things by the wrong name, inadvertently opened a portal to Hell and made it possible for legions of demonic entities come charging into our previously-lovely dimension of reality.

While we are on the subject, there is an item on my "bucket list" which I would like to attend to, if nobody minds. It's just a little something I wanted to pass along to all the nice folks who work at the supercollider. I figure I may as well do it now, while I am still temporarily above ground and have the chance to.

It is said that the soul of wit is brevity. If that is true, I would guess that this next part ought to be considered an instant classic:

"On behalf of all Earth's inhabitants, I bring you a message! It consists of only two words. Let's see if you can guess what they are."

That is all.

CHAPTER 17: REWRITING THE PAST

"Every record has been destroyed or falsified, every book rewritten, every picture has been repainted, every statue and street building has been renamed, every date has been altered. And the process is continuing day by day and minute by minute. History has stopped. Nothing exists except an endless present in which the Party is always right."
— *George Orwell*

"If I had it all again, I'd change it all."
--Bush

In the previous volume ("Alien Contact: The Difficult Truth") I touched upon the issue of technology which makes it possible for those who possess it to travel both forward and backward through time. From our parochial, Earthbound view, time travel technology is for many people a bridge too far. It is more than they are willing to accept, and those who speak of it in a meaningful way are immediately in danger of being disregarded as being either idiots or frauds.

I understand this well...and yet, at the same time, it is necessary to keep in mind the fact that we are not dealing here with conventional human scientific limitations. We are dealing with beings whose level of scientific expertise—and the commensurate capabilities which go along with it—are vastly superior to our own.

While time travel may be nothing more than a theoretical possibility as far as our mainstream scientists are concerned, it is something which has been turned into a practical and usable technology by beings who are far more advanced than we can imagine. Attempting to limit such beings by imposing upon them the limitations of our own scientific progress is a tremendous mistake, and one we cannot afford to make.

As we are constantly reminded when studying the phenomenon of alien contact, it is necessary to free our minds from the constraints which are naturally imposed by thinking only in terms of our conventional, non-classified capabilities. Without breaking free of the box, it is impossible for anyone to come to an accurate understanding of either the civilizations we have determined to study, or the situation humanity currently finds itself in.

Does the idea of time travel prove difficult for the average person to accept? Yes, most of the time it does. I am the first to admit that it may have been more than I was willing to accept myself, had I not experienced it firsthand. When I make the statement that time travel is not only possible but is being utilized on a regular basis by both alien races and our own black ops military groups, I do so from a position of knowledge, not speculation.

The aspect of time travel I would like to discuss in this chapter is the idea, which comes from multiple inside sources who have chosen to become whistle-blowers, that certain alien groups have made a habit of traveling back into our past. From there, it is reported, they have manipulated events in a manner which results in changing our history in a way which proves favorable to them. I'm sure we can all agree that if we were in their position, we would not hesitate to do the same thing.

I have been informed that the United States, although originally unaware that intervention in and manipulation of our past by alien beings was taking place, eventually became aware of it. At that point, I'm told, the powers that be insisted that such activities immediately be halted.

I have also been made to understand that, once they had gained the ability to travel through time themselves, units of the American black operations military forces were sent back into the past on a parallel mission. Once there, they were to alter past events in ways which would bring advantage to the United States and be manifested in our current time. Apparently, the American military felt that, although alien races should be forbidden to alter our past, no such restriction should apply to them.

Whether such activities on the part of the U.S. military are still ongoing is not known to me. I presume that they would continue to manipulate events in the past for as long as they felt it was useful or necessary to do so. If it is no longer being done, it is almost certainly because they feel they have already achieved all they can. If that is *not* the case—if they still feel there are advantages to be gained for the military by tweaking the past—then, in my opinion, we should operate under the presumption that they are still at it.

Careful consideration of the ideas contained within this chapter should result in a new appreciation of the possibilities and potentials which are inherent to possessing the ability to travel through time. Those who control the past also control the present, and those who control the present control the future.

When world events and power structures are viewed from this perspective, we can see even more clearly the extent to which the entire human race has been made into pawns, forced to participate in a game they are not even aware of. We live out the course of our lives, never suspecting a thing, even as we are pushed around the board by the unseen hands of invisible masters who are not in the least concerned with our best interests or welfare.

This is the state of the world today, brothers and sisters, and perhaps it has always been so. I see no compelling reason to believe otherwise. When the full extent of the power wielded by these shadowy figures and alien intruders is considered, it is difficult to find any scenario which makes me feel comfortable.

We are encouraged to believe in self-determination. We are taught to think that the ability to shape the future lies within our hands. In the real world, however, our power to shape the future is sharply limited and defined by those who truly *can*—and, as noted, this group is made up of those who control both the past and the present. This results in a situation which is extremely dangerous to us and is surely among the most urgent and high-priority problems we face.

If these groups maintain their current level of influence, we will be unable to control either our future or, ultimately, our fate. Even our most powerful military forces—and their commanders—will be compromised—and potentially controlled—by aggressive, intrusive, decidedly hostile powers.

Please believe me when I say that being the bearer of such news is not something I ever aspired to and is distinctly *not* something I enjoy. The fact that it has somehow become both my job and my duty, something I am unavoidably called to do regardless of my personal wishes, is a matter of destiny it seems. If I could avoid spreading such news without placing others in danger, I would certainly do so.

It is an unfortunate fact, however, that those who misunderstand the nature and scope of the dangers we face are little more than helpless, stationary targets who can do almost nothing to save themselves. I understand clearly that few of those people will ever actually read this manuscript, and that most of those who do will fail to heed the call to action it contains. That is beyond my ability to change—it is the way of the world, unfortunately.

For the few who will read these words and choose to do all within their power to help save what is left of our society and our world, I have dedicated not only this series of books but my life. It is far beyond my power to predict what shape the future will take, or to affect it in any significant way…but let it never be said that I stood silently by and did nothing while my people were in danger.

Circumstances have dictated that I deliver these warnings and sound the alarm. What others may decide to do with this information is out of my hands. I can only do my best and hope you will choose wisely.

CHAPTER 18: INTERSTELLAR HYBRIDS

"I am quite sure that our views on evolution would be very different had biologists studied genetics and natural selection before and not after most of them were convinced that evolution had occurred."
— *John B. Haldane*

"That's me in the corner,
That's me in the spotlight
Changing my religion
Trying to keep up with you
And I don't know if I can do it,
I don't know if I can do it
Oh no, I said too much
I haven't said enough!"
--R.E.M.

In the first volume of this series, a connection of mine I referred to as "The Colonel" stated that, according to alien medical technology "which far outstrips our own abilities", DNA is imbued with a timer, a sort of natural time bomb, which will eventually expire. When this happens, the DNA has essentially become worn out, it has lost its versatility over time.

This could be a result of a population interbreeding over such an extended period that one person's DNA is much like another's, to the extent that any offspring which are produced will not be different from either of the parents in any meaningful way. It could, on the other hand, simply be a natural timer that expires at a certain point regardless of that.

In any case, says this source, when the timer does expire a race finds it increasingly difficult to reproduce naturally. Eventually, natural reproduction becomes impossible. I do not know exactly how or why this is the case, but the source in question is in a position which allows him to know what he is talking about, so I will presume for now that what he said on this issue is correct.

When natural reproduction becomes impossible, a species soon becomes extinct. According to my source, this is something which is seen as a natural occurrence by ALL the alien races we are aware of—

they consider it to be "the way things ought to be". They apparently think of it as nature's way of ensuring that life itself remains versatile and ever-changing as the old life-forms eventually die off and make room for the new to take their place.

In the case of intelligent races, however, the inability to reproduce naturally does not automatically mean extinction for an entire race. If they have developed the ability to reproduce by cloning as well as by natural methods, the lifespan of a race can be extended by a considerable amount by producing clones of individuals to replace the originals.

Cloning, the Colonel said, is an imperfect technology in the sense that with every subsequent generation of clones, a small amount of the original genetic information becomes lost in the process. The clone will then be slightly inferior to its original cell donor. It will, among other things, tend to be weaker overall and to have a shorter lifespan when compared to the original cell donor.

Over time, as the lifespan continues to decrease, and the succeeding generations of clones become generally weaker and less efficient than the originals were, there comes a point where the process of cloning itself becomes useless in practical terms. Clones produced at this point will have extremely short lives and lack the vital energy required for successfully managing even those drastically-shortened lives. Extinction of the race will soon follow, when this occurs. It will no longer be preventable even by artificial methods and the race will die out.

We can see from the fossil record here on Earth that most species which have ever existed here have, at some point, become extinct. Reasons for their extinction are theorized in some cases, such as the "dinosaur killer' meteor impact millions of years ago, but the fact is that in most cases there is no compelling reason which is known to be the cause of the extinction of most organisms which are no longer here.

It is presumed that conditions changed somehow, and the organism was unable to adapt to the new conditions and therefore died off. While this is a reasonable theory, it is not something which is known for certain or which can be scientifically proven in most cases. Species appear, they flourish for a time, and then they become extinct—the same process occurs time after time throughout the fossil record, and there is no apparent reason why this should be the case.

The reason I mention this here is that our universe is a very old place. Our sun is considered a "third-generation" star, meaning that two generations of other stars lived out their lives and died before the processes which resulted in our sun being formed even began. The time required for its formation alone could have extended to several billion years. The process of the planets in our solar system forming and then eventually producing life probably required several billion more years to occur.

Although humanity as we know it appears to be very young in cosmological terms, the universe itself is not. We cannot even be certain that our sun is a third-generation star—for all we know, there could have been many generations of stars which came before it.

If we presume, for arguments sake, that the first generation of stars which came into existence were composed of pure hydrogen because no heavier elements yet existed, those heavier elements would have eventually been produced by that original generation of stars. As those stars consumed the last of their hydrogen fuel, converting it to helium, the helium would then have been ignited and fused, eventually being transmuted into the next heaviest element.

This process is repeated time after time, one element after another, until a point is reached where the star's gravity and the immense power of the fusion-heated elements blasting outward from the star's surface became imbalanced. At that point, the remnants of the star would either collapse into a primeval black hole or explode in a nova or supernova event, thereby casting all the heavy elements it had produced over its lifetime out into the universe with great force.

These elements, forged in the heart of dying stars, were the raw material from which planets, moons, asteroids, comets and life itself were eventually formed. Every molecule in our bodies was originally created inside the heart of a dying sun. Ejected outward, from the star which had formed them, they drifted through space for millions of years, eventually coalescing under their own gravity, colliding and sticking together, forming objects of increasingly greater mass. They ultimately became the raw material from which the next generation of stars—and the planets which orbited them—were formed.

But this is physics. We are not here to enter a detailed discussion of physics, but to discuss the topic of alien contact. So why does any of this matter? How is it relevant to the topic at hand?

It is relevant because of this idea, one which is almost never mentioned in ufology circles. There has been more than plenty of time for other intelligent races to have come into being, to have developed highly advanced civilizations and then to have ultimately become extinct due to the expiration of a DNA timer. There is, therefore, no reason for us to presume that any of the races we now see visiting our world from elsewhere are necessarily being seen in their original form.

There is another way to prevent extinction other than cloning, you see: hybridization. By successfully combining the DNA of two individual intelligent races with genetic engineering technology, a hybrid is produced which has DNA unlike either of the parents. Its DNA will necessarily be a combination of the two original parent races, a new race unto itself which was produced from two others.

This new race, when created, will not be bound by the DNA timers of either of its parent races. It will have a brand-new DNA timer, with its own term and projected usefulness. This new race, which will have characteristics of each of its parents, but which will not truly be a part of either of their races, can survive for millions—or, perhaps, even billions—of years on its own, because rather than starting off with a timer that has reached its final stage, it will have a timer which has only begun.

Rather than having DNA which has become useless and worn-out, it will have young, strong, vital DNA which was the result of a wholesale transfer of new genetics to the new organism. Rather than becoming extinct, the two original races which produced the DNA used in the hybrid will exist on into the future in a changed form. Each of them will now be half "alien", having taken on half of their DNA from the other, but some part of them will indeed be projected forward through time for millions of additional years.

The alternative is to one day end up with clones which are so weak that they themselves can no longer be used to generate clones from. When this occurs, extinction must soon follow.

One can easily see this general principle at work by cloning plants, a straightforward process which is often used on plants which carry genetics one wishes to preserve. It might involve, for example, an especially strong and vital tree, one which you for some reason wish to plant many copies of.

The original clone of such a tree or plant will be strong and vital, just as the original was. This plant can then be used to make additional clones from, and those clones can in turn be used to make even more, in their time.

With successive generations of clones, however, one will encounter a problem. Each successive generation, though it will still produce clones with the same genetic characteristics the original plant had, will become slightly weaker than the original. They will not grow as large, or as fast. They will not be as healthy or robust.

Eventually, after a certain number of generations has passed, the clones which are made from that generation will be so weak that in practical terms they will be all but useless. Rather than growing tall and strong, they will be sickly and struggle just to survive.

The only solution is to create a hybrid from an earlier generation of clones before the clones reach this point. A similarly strong and excellent plant is found, one which is not genetically related to the clones. This plant is used to pollinate the clones while they are still in a relatively good condition, meaning they have not been cloned for so many generations that they have become weak.

The clone, fertilized by the pollen from this second strong plant and still carrying the characteristics which made the original plant so desirable, will produce seeds. From those seeds will come a hybrid plant. This hybrid will benefit from something known as "hybrid vigor"—the combination of two sets of DNA which were completely unrelated to each other will result in a plant which is stronger than either of its parents. It will sprout sooner, grow faster, grow larger and produce better fruit than either of its parents did.

This is obviously a far better fate than is becoming extinct.

It is quite possible that intelligent life evolved in other places long before our sun was even formed. It is therefore logical to presume that many of the alien races we now find ourselves dealing with have, at some point, reached the end of their original DNA timers. They have been forced to hybridize themselves with other intelligent races to avoid extinction.

This would mean that the at least some of the aliens we are familiar with today are not in their original forms but are themselves hybrids. It could even be the case that this applies to literally *all* of them.

Depending upon how old the universe is and exactly when intelligent live evolved, such hybridization could in theory have taken place more than once. For all we know, we could be dealing with hybrids who were created using the DNA from hybrids which were themselves created using the DNA from hybrids which were themselves created using the DNA from the original incarnation of some intelligent life-forms.

This provides an answer to the quite reasonable question of why so many of the alien races we are aware of look so much like each other and to humans. We would not expect such a thing to occur naturally. If each intelligent race was the product of a completely independent evolutionary chain, we would presume that there would be wild variations in form between different intelligent races.

It may well be the case that there are advantages conferred to an upright, bipedal form—one with two arms, two legs, etc.—which result in the selection of such a form on many different worlds and which could eventually result from multiple evolutionary pathways. But to have so many of them end up with such a strong resemblance to each other—and to homo sapiens—due to completely independent evolutionary processes would be so unlikely that the odds of it occurring naturally would be vanishingly small.

It also seems to be the case that some of the alien races we are familiar with are no longer located at their planet of origin. Their original home appears to have become unsuitable to sustain them for some reason. Perhaps all its natural resources have been used up over time, or it has become too polluted to serve as their home any longer. It may be that the star their original home worlds orbited may have passed out of its main sequence, meaning that it has changed its form over time and the planets where they originated were no longer capable of sustaining life.

This will eventually happen to our own sun, too. Once the elements that it utilizes for its fusion process are exhausted, it will become what is known as a "red giant".

When this occurs, the sun will cool down and greatly expand in size. It will become so large that it will eventually enclose the orbital space of Earth. This will be, quite literally, the end of the world.

If humanity still exists when this point in time draws near, there will be no alternative but to find another world to call our home. We will be forced to locate and inhabit a world in some other solar system, a place with a star that has not yet passed out of its main sequence.

I believe that this is what has occurred to some of the alien races which are either visiting our world now or have done so in the past. If so, it serves as another illustration of the great antiquity of these alien civilizations and yet another reminder of just what we are faced with in our contacts and dealings with them. While humanity's civilization is measured in mere thousands of years, we are almost certainly dealing with alien civilizations whose history stretches millions—and, in some cases, billions—of years into the past.

On Earth, a nation whose technology falls behind that of its enemies by as little as a few decades will normally be incapable of winning a war against them. Imagine the situation if they happened to find themselves relying on technology which was ten million years behind that of their opponents! The only possible outcome would be an extremely short war, resulting in the complete destruction of the more primitive nation and the subjugation or elimination of its citizens.

In terms of human technology vs. the technology available to alien races, there is in my view no possible comparison. We are dealing with alien civilizations which appear to have had the technology to travel to Earth long before humans even existed. Moreover, they did so using not only interstellar travel, but *interdimensional* travel as well. Millions of years ago—and, for all we know, possibly tens or even *hundreds of millions* of years ago—these beings had mastered both interstellar and interdimensional travel.

If it comes to a violent confrontation, our chances of victory would appear to be less than zero, if such a thing is possible. In terms of defending our world against a subtler assault, such as the one which appears to be taking place now, we would similarly be at such an enormous disadvantage that it does not even bear thinking about.

We cannot possibly hope to hold our own, or even remotely come close to doing so, if conflict arises between ourselves and any of these ancient civilizations which apparently surround us on all sides. In the end, our existence depends upon the willingness of alien races—none of which we should presume owes us anything, including empathy or compassion—to allow us to survive.

If, at any point, they choose to revoke that permission, there is almost certainly very little we could do to save ourselves. Our good behavior, and the care and wisdom with which we comport ourselves when dealing with these alien civilizations, will very likely be the deciding factor in whether humanity itself continues to exist. This, it seems to me, is an idea which is difficult to argue with.

To imagine that we may someday become their technological equals is, in my opinion, nothing but wishful thinking. How, after all, does one close a technology gap of ten million years? By adding a night-shift at the research and development facilities, and promising a handsome bonus to whomever manages to construct a fully-operational hyper-dimensional battle cruiser?

Hardly. It isn't possible to close a gap like that—and while the gap remains, it isn't possible for us to be considered their equals no matter what the arena may be. We are not co-citizens of some galactic brotherhood or collective—we are ants, and they own the ant farm. This is an objective reality which most humans will surely find extremely difficult to accept and to live with. But what, specifically, is the alternative?

There isn't one, as best I can tell. Reality does not always conform to our wishes, and objective truths have nothing to do with what we might feel is "right" or "fair".

We do not control the situation regarding the agenda of alien races or civilizations. All we can hope to control is our own interactions with those races, and it is those interactions upon which our future will depend.

It is surely in our collective best interest to ensure that our relationships with various alien powers and principalities is one which all concerned can live with. We are involved in a situation which we have no reason to presume is likely to offer us any second chances. It is of critical importance, therefore, that we find and utilize the optimal solution on our first try.

CHAPTER 19: FULL PUBLIC DISCLOSURE

"The truth will set you free, but first it will make you miserable."
-- James A. Garfield

"The danger on the rocks is surely past
Still I remain tied to the mast
Could it be that I have found my home at last?
Home at last..."
-- Steely Dan

Many people truly believe that some type of full public disclosure on the part of the United States government is drawing near. It is my opinion that these people, sincere and well-intentioned though they may be, have either not thought the matter through carefully enough or do not have access to enough information to understand the situation as it appears from the government's point of view.

If they had, it seems to me that they would have eventually arrived at one inescapable conclusion: the government of the United States has no intention of proceeding with any type of full public disclosure until and unless it has no other option available to it. Furthermore, it must be noted clearly that if such a disclosure were to occur it would without question be less than complete, nowhere approaching honest and filled as full as possible of disinformation and lies.

It is instructive to consider the question of full public disclosure from the viewpoint of those who ultimately control access to the full range of information associated with alien contact. When examined from their point of view, it looks very different than it does to the average member of the public.

The first problem to be dealt with will be the issue of who exactly will deliver such a disclosure to the nation and the world. Clearly, no less a personage than the President would be acceptable or credible to the public at large were they to make such an announcement. No other official carries enough weight to be able to pull it off successfully, and have his statements accepted as fact by the civilian population.

Already we encounter a rather large problem: The President of the United States does not hold a high enough security clearance to be allowed access to complete and unfiltered information about alien contact. Whether that is justifiable or wise on legal or moral grounds is a matter which can be speculated on—nevertheless, it is a fact of life and one we must live with.

This lack of all-encompassing clearance by the President means that he will necessarily be informed of only part of the story. Many things—and certainly many extremely important and critical things will be numbered among them—will be withheld from him by those who brief him on the issue.

The attitude of the military/intelligence services is this: In a few years the President will be out of office, once again a civilian, and will never again give an order to any military unit or decide any policy issue. The military, however, will remain...and so will the extra-terrestrials and any ongoing projects which are associated with them. These projects, many of which are long term and horrendously expensive, cannot be allowed to be jeopardized by the decision of some do-gooder President—no matter whether he happens to be Commander-in-Chief or not.

That being the case, any information or any projects associated with extra-terrestrials which the military considers to be of high importance will be kept secret from the President. They will not be mentioned in any Presidential briefings and the President will have no reason to suspect such projects or information even exists.

This puts the President in the position of being unable to make well-informed decisions regarding anything to do with the overall extra-terrestrial situation. It also makes it impossible for him to accomplish any type of full public disclosure on his own, even if he wants to do so.

Without access to the full range of information which is available, there can *be* no full public disclosure. The most that could occur would be partial disclosure. Partial disclosure, I think it is safe to say, is a solution which will be found satisfactory by exactly *nobody* on either side.

Let's set that issue aside for the moment. For the time being, let's presume that the President *was* given full access to all relevant information. Let us further stipulate that all parties involved were guaranteed full amnesty in exchange for full public disclosure of the facts with nothing held back. We must now examine—from the viewpoint of the government—some of the information which would presumably be made public if disclosure were to take place.

While absorbing and considering the remainder of this chapter, I would ask the reader to keep a couple things in mind:

Firstly, I have gone to great lengths to ensure that the statements which are to follow are both true and accurate. I am not, however, naïve enough to presume that this is always the case. In the present instance, for example, I do not flatter myself by believing that I was somehow able to nail it perfectly on my first try and got everything exactly right.

I proceed under the presumption that it is virtually certain that I have made some mistakes on this, despite my best efforts to avoid doing so. I do not possess anything even remotely close to all the data regarding this subject which exists, nor have I been given a security clearance which would make it possible for me to access that closely-held information.

It is also not a complete list, for the same reasons. There are undoubtedly many additional reasons that I am unaware of. There are also quite a few I know of, which it didn't seem necessary to include here.

With that in mind, I strongly recommend that the reader adopts the same attitude toward this material as I have taken myself. Do not assume that I got all the minute details right, because it is virtually certain that this is not the case.

Having freely admitted that I cannot do something which would be all but impossible for a person in my position to do, the best I can hope for in the real world is to come reasonably close to it. That is a somewhat different matter. That is something I might be able to pull off.

I feel confident that I can provide you with far more accurate statements than will be required to drive the point home, and to do so in a way which I think will be difficult to logically dispute. We shall soon find out.

What I would ask the reader to do during the remainder of this chapter is to imagine that the day of full public disclosure has finally arrived. Imagine, if you will, that the President of the United States is holding a world-wide press conference, during which the truth will at last be disclosed to the public.

As he steps to the podium, the entire world holds its breath. Their attention is concentrated like never in history on the individual whom destiny has chosen to put center-stage at this most critical moment. He clears his throat, takes a drink of water from the nearby glass, and faces the sea of cameras which is assembled in front of him. His face displays the proper gravitas and somber attitude we would expect of him as he glances down to his notes and then begins to speak.

His opening remarks make little difference to us here. Imagine them to be whatever you like. What matters is that, during his speech, he will be required to include the following statements in some form:

"If you are a citizen of the industrialized world, and particularly if you happen to be a citizen of the United States of America, you have been targeted by a long-term, extremely comprehensive, incredibly well-planned, devious and subtle form of psychological warfare. Your cognitive abilities have been steadily and efficiently eroded by a variety of methods, including the fluoridation of your drinking water, educational institutions which have grievously misinformed you about pretty much everything, and other things besides.

"Your television sets have been used to brainwash, hypnotize and desensitize you, and the news you are given has been nothing more than carefully-written propaganda and subtle manipulation designed to control your actions, beliefs, words and even your thoughts. This type of mind control is so incredibly effective that it is impossible for anyone who watches television to avoid or overcome.

"As you listen to me now, I tell you that the thoughts in your heads—which you believe to be your own—are in fact not your own at all. They are ours. They have been carefully and methodically implanted into your consciousness over years and decades, and you have been caused to think, say and do exactly what your unseen masters have intended you to think, say and do.

"Who was it that decided what those things would be? Was it people who guard your best interests and protect your safety?

168

"Hardly. The people who decided these things are completely unconcerned with your safety and your best interests. Some of them control vast financial empires, others direct super-secret intelligence agencies and still others hold the highest positions of command in our secret, unacknowledged military special operations forces.

"Their interests and yours have nothing in common at all…and their interests have been served, not yours. You will probably be surprised to learn that almost all of them are black magicians, Satanists, and that they are involved in such things as human sacrifice, pedophilia, murder, coercion, secret control of both the military and the government, and many other things besides.

"We have only begun to scratch the surface so far, in terms of the full public disclosure that so many of you wanted so badly. And believe me, my fellow Americans, this is the least of it. This is the good news! Are you still happy that the day of disclosure has arrived at last? Or has it, even here as I speak the first words of a long presentation, already lost some of its shine?

"It makes little difference, because as I said this is nothing more than the tiniest tip of the iceberg. We have far to go, and much remains to be said. Even so, I can guarantee that right now, across the country, there are tens of millions of American citizens who are pouring themselves a stiff drink—and most of those drinks will be double shots.

"Extra-terrestrial entities have been present on and around Earth throughout the entire history of humanity.

"It seems evident that various extra-terrestrial races have intentionally manipulated and interfered with human societies, governments and religions, denying us the ability to make our own choices without interference from the outside. It also seems quite clear that manipulation of this type is ongoing and continues to this day.

"After spending uncounted trillions of dollars to produce and support what is by far the most powerful and intimidating military force in history, the truth is that we do not even control our own airspace, we never have, and we probably never will.

169

"The technology available to these extra-terrestrial beings is advanced far beyond our own capabilities. We do not have the ability to eliminate their technological advantage and pull even with them, much less surpass them. Not now, not a hundred years from now, not ever.

"Extra-terrestrial beings are currently engaged in a program which involves abducting large numbers of our citizens from their homes in the middle of the night and performing various medical procedures on them against their will. The technological advantage enjoyed by the ET's is such that our military is completely incapable of stopping these abductions or protecting the safety of the citizens who are victimized by them.

"An American President has agreed to a secret treaty with a group of these visiting entities in which we basically agreed to trade our citizens to the aliens in exchange for certain alien technologies, most of which have never been made available to the public or even admitted to. The treaty, though unconstitutional and entirely illegal according to our laws, has bound us for most of a century at present.

"The U.S. government has, largely because of the treaty, come into possession of numerous highly advanced technologies and capabilities. Among the technologies which it currently possesses are the secret to "zero-point" energy—energy which, though technically not quite free, is so inexpensive that it would be affordable to everyone and would immediately end our dependence on ever-more-expensive oil and petroleum products.

"Also, among the technologies the government has acquired are the cures to virtually all human diseases including cancer. These technologies have been intentionally withheld from the public, even though they would clearly provide immediate benefits and long-lasting positive change if they were to be released.

"The interests of the public at large were in this instance—as is virtually always the case when it comes to anything which deals with extra-terrestrial contact—deemed irrelevant and unimportant when balanced against the government's perceived need to strengthen and solidify the position of individuals within it who hold enormous and unwarranted power over the future of both our nation and the other nations of Earth as well.

"I asked one of our General officers why the cures for such widespread maladies as cancer were withheld, rather than being shared with us, as one might expect would be done. His response to me was (and these were his exact words): "Because fuck you…Mr. President."

"When they told me that certain of our high-level military officials would only be willing to speak the truth after having received an official guarantee of amnesty from prosecution, they apparently weren't kidding.

"If any of these extra-terrestrials should turn out to be overtly hostile, our military forces will not provide us with any realistic type of defensive capabilities and will be unable to defend either the territorial integrity of the United States or the safety of those who live here.

"Humanity does not hold a position at the top of the food chain, as far as the greater community of interstellar, inter-dimensional and Earth-born visitors are concerned. At least one alien race, known as the Greys, ingests adrenaline-laced human blood regularly and with great gusto. This alien race is considered hostile, parasitic, dishonest, manipulative and invasive, and has demonstrated all these characteristics repeatedly over a period of many decades.

"Our military does not have the ability to adequately defend our nation against them or to force them to go away. Because of that fact, this is the very group of extra-terrestrials with which our government signed that secret, illegal treaty in 1954.

"It appears quite unlikely that the Greys are the only race of alien beings which make a habit of utilizing humans as a form of food and/or sustenance. I understand that none who hear this will be either pleased or encouraged to receive the news. There is, however, nothing we can do to change their dining habits, and precious little we can do to prevent the 'harvesting' of human individuals whenever they choose to do so.

171

"As you may have inferred from my previous statements, humans are seen by several the alien races which do or have visited our world as a renewable resource, to be harvested as necessary. We are deserving of no protection or moral rights as far as they are concerned.

"Before either beginning negotiations or signing the treaty agreement, the United States was offered a treaty by a different alien race, one which was non-violent and promised to assist us in solving our global problems. Because that race refused to provide us with military technology as part of the agreement, their offer was turned down in favor of trying to reach some type of agreement with an alien race which was even then considered to be hostile and invasive, but which would provide us with militarily-useful technologies which could be used against other nations of the Earth.

"Several thousand American citizens have been murdered in cold blood by agents of the government for having had the bad fortune to somehow learn about the truth of an extra-terrestrial presence and/or having the bad judgment to speak about it publicly. These individuals committed no crimes, overstepped no moral limitations and did nothing to deserve acts of first-degree murder to be carried out against them.

"The concern about keeping the truth about alien contract secret from the public outweighed any inherent right to freedom or to life they may have had, and it was therefore decided that they would be terminated. If you have ever wondered how concerned the government of the United States is about justice, freedom of speech or providing a safe environment where its citizens can live free from fear of reprisals by tyrannical governmental forces, you may now have the answers you were seeking.

"Control of alien-related technology and projects has long ago passed out of the hands of the elected members of the federal government and those they command within the military. These things are now controlled by a very small group of corporate elitists who are willing to use any means or methods to protect their exclusive access to knowledge and technologies which bring them huge profits and provide them with enormous advantages over other members of society.

"It is them, not members of the United States government, who make the decisions which control policies and actions related to alien contact and alien technology. These people are loyal only to themselves and their continued access to unwarranted levels of power and unfairly-gained financial fortunes. Simply put: we couldn't stop them even if we wanted to.

"Elements of United States special operations forces have for many years been engaged in the abduction of American citizens, who were utilized for their own purposes without regard to their well-being or the constitutional safeguards guaranteed them. The majority of so-called "alien abductions" are in fact military abductions which utilize alien tech-based craft and technologies. They allow our agents to walk through walls or make themselves appear to be extra-terrestrials. This is done to deflect the blame from our troops should anyone happen to retain memories of the abductions.

"One of the purposes for military abductions are the forced breeding of females to impregnate them with what will eventually be our own group of alien-human hybrids, returning in 90 days to steal the fetus from the mother's womb and raise it to term in a vat. Another purpose is to use the victims as unwilling remote viewers for purposes of clandestine spying. Yet another is to use them for slave labor, without feeding them at all. When they are about to expire from abuse and starvation, time travel technology is utilized to return them to their homes as though no time had passed, and they had never been gone at all.

"Though it was cleverly hidden from view in such a way that you surely did not notice it and were never aware such a thing had occurred, there is a little surprise which awaits the American people when it comes to their money. We didn't think you would mind, if you had no idea it was even being done. So, we took the profile of a Grey alien, complete with a circle of stars where the so-called "third eye" is located to illustrate that they are telepathic entities and…well, we put it on all your dollar bills.

173

"The presence of an extra-terrestrial on your currency will naturally lead many of you to presume that your government has in fact been co-opted by an alien nation and is now controlled by them. You will assume that your leaders are nothing more than puppets beholden to alien masters. That is, unfortunately, largely true.

"The design itself has not been modified since the year 1929. Make of that what you will. Perhaps the United States government was communicating with the Greys earlier than is commonly believed to be the case. Perhaps the image of a Grey alien does not actually appear on our dollar bills. We know that nothing appears on our currency by accident or mistake, however please feel free to form your own opinion about this, as seems best to you.

"It is also true that Earth is being silently invaded by these alien entities, who have every intention of decimating the human population, eliminating literally billions of people until the remaining humans are at a low enough number that they will be easy to control.

"At that point they will be enslaved, according to their plan, and they will emerge as the new masters of what was once our world. You may perhaps take comfort in the fact that we do not believe there is any reason to think that the Greys can make this happen.

"The reason they may not be able to complete their conquest of Earth is because there are several other ET races which also apparently intend to dominate this planet and emerge as its new owners and masters. We do not know which of them will end up being successful in the end.

"Don't misunderstand: if you are among those who survive long enough to experience it personally, you will become the slaves of some hostile alien race before all is said and done—we just aren't sure which it will be. Hopefully one that doesn't eat humans alive or bathe in their blood, as I'm sure you'll agree. But we will just have to wait and see how it all works out before we can know for certain.

"The true purpose of NASA has never been to explore space or to keep the public informed about the latest developments related to such exploration. NASA is and was always intended to be a hugely-expensive tool of propaganda and a source of intentional disinformation which was seen by the public as being inherently honest and trustworthy.

"The true purpose of NASA is and has always been to prevent members of the public from learning the truth about what is really taking place, both on Earth and in our solar system, and to make them believe they had no reason to doubt NASA's official pronouncements or to look beyond them for further information. Its secondary purpose is and has always been to provide a convenient place to hide enormous amounts of funding, which was channeled off into black projects which the public was not informed about and did not gain any significant benefits from.

"SETI, the Search for Extra-Terrestrial Intelligence, which utilizes a network of enormous radio telescopes to scan the heavens in search of some type of signal or communication from distant alien civilizations, is and has always been covertly controlled by military personnel. We would of course never allow such a powerful technology to be operated and controlled by any group of civilians.

"The inherent risk that they might receive some type of interstellar communication and then inform the media is a risk we are absolutely unwilling to accept. For that reason, all such potentially problematic circumstances are prevented from occurring. Any such signals which may be received are immediately relayed to the relevant military officials, who then take measures to ensure that the press is never made aware of them.

"At various times during its existence, SETI has indeed received such communications from civilizations which are very distant from us. These include some which originated in galaxies other than the

Milky Way and have made their way across intergalactic space for millions of years before reaching us.

"The actual space program being carried out by the United States has never been planned, directed or controlled by anyone employed by NASA. It is, and always has been, a program which falls completely under the authority of the United States military. Primarily, this is the responsibility of a hitherto-secret branch of the service known as the Space Force.

"Its budget, goals, missions, discoveries, programs and accomplishments have been completely hidden from both the public and the press, and its existence denied by all governmental offices and personnel. The Secret Space Program has for many years availed itself of a wide spectrum of alien-based technologies, for a multitude of purposes. These purposes include, but are by no means limited to, exotic methods of propulsion for our craft which allow us to travel much faster, and much farther, than members of the public believed were possible.

"Among the projects which were undertaken by this secret, unacknowledged space program was the establishment of permanent manned, underground bases on the surfaces of both our moon and the planet Mars. These bases have been operational for decades and remain fully-staffed and operational at present.   *

"In addition to black ops military personnel who are assigned there, the Mars colony also makes use of a human labor force which consists of individuals who are not connected with the military and were transported to Mars involuntarily, illegally and forcefully by members of our covert military forces.

*All the information regarding the planet Mars, including those which refer to its forests, atmosphere, alien inhabitants and the secret underground human base which is located there, would presumably have to be included as part of any Presidential statement which provided full disclosure to the public. I have not repeated them here, for the sake of brevity.

"Once on Mars, they were relegated to slavery and utilized as a permanent contingent of unpaid laborers within the mines we operate there. This is not technically illegal, because there are no anti-slavery prohibitions in effect on the planet Mars—or, for that matter, laws of any type, other than those which are written and enforced by the alien landlords. We are not restricted or prevented from owning or utilizing human slaves in our operations there, and we have taken full advantage of that by using them constantly and spending as little as possible on their food, water and shelter needs.

"Our extra-terrestrial colonization program by now also includes various planets located in distant solar systems, which we can reach by making use of exotic propulsion systems which are based upon alien technology and require minerals which cannot be produced here on Earth to function. As with the lunar and Mars bases, we have colonized these worlds with the knowledge and permission of the aliens which control both our world and our solar system.

"Those who are in power have, entirely at your expense, constructed a massive system of underground bases and cities linked together by futuristic mag-lev subways which can attain speeds far greater than the speed of sound. All these bases and cities are fully supplied and well-stocked with all the necessities and are self-sufficient and self-sustaining. They can ensure the survival of those who inhabit them for literally hundreds of years, completely sealed off from the above-ground world.

"They are also equipped with a plethora of high-tech weaponry and literally billions of rounds of highly lethal ammunition, also purchased at your expense. The entrances to these facilities are all highly camouflaged and protected by the latest technology, including holographic projectors which make the entrances appear to be nothing more than random uninteresting hillsides.

"These entrances consist of the most well-designed and secure blast doors ever built by humans. They are identical to the blast doors which protect the entrance to NORAD Supreme Headquarters at

Cheyenne Mountain, and are specifically designed to be able to survive anything short of a direct hit with a nuclear weapon.

"Those who happen to be somehow prevented from entering these underground facilities and finds themselves outside the blast doors after they have been sealed shut will, of course, be completely unable to force them open. Even if such a thing were possible—which it manifestly is *not!* —anyone who managed to bypass those doors would be obliterated in very short order by the elite special-ops forces who would be waiting for them on the other side.

"The roster of geniuses we employed to design, construct and provision this vast underground network—which is far and away the most high-tech survival shelter in all human history—left nothing to chance. They truly thought of everything.

"You, who were told nothing at all about any of this, were kind enough to finance the whole thing with your tax dollars. Assisted, of course, by a national debt burden so great that it will soon become impossible to pay even the interest owed on it, to say nothing of being able to pay down the principle itself.

"When everything goes bad, the elite of this world will be taken into these underground facilities to ensure their own safety and security. Others will be taken as well, people who are deemed to be necessary to their long-term health and survival and who will be forced to work for them under threat of elimination.

"You are not among those people and will be left on the surface world to fend for yourselves against threats so serious that the rich and powerful found it necessary to seal themselves in underground bunkers to guard against them. None of you are expected to survive.

"Those who will be making use of bunkers and will be safe and secure, literally miles below any potential threat which may be present on the surface world, have pointed out—quite rightly—that this will be *your* problem, not *theirs*.

"By now there is of course far more than ample evidence to convict your military, political and economic leaders of high treason and war crimes and more than sufficient justification to have them all executed. We are fully aware of that.

"Those of you who have studied probabilities may well be able at this point to accurately estimate the chances that their little disclosure petitions would realistically have of forcing or convincing their government to proceed with full public disclosure. For those of you who are less mathematically inclined, I will be happy to provide that number to you in the interest of full public disclosure. That number is—and always has been--exactly zero percent.

"All your efforts in that regard have been entirely pointless. There has never been any chance whatsoever that any of them would succeed or even be seriously considered for so much as a single second by those who control access to that information. And we are not done here yet. There is more to come…and I am sorry to inform you that none of it is any better than the things we have already discussed.

"There is more which could be said here. The list could be extended quite a bit. But it seems appropriate to group many of the other matters together here in a single entry: It is not true that all ET's are hostile. Some are friendly.

"Due to our obligations under the treaty we signed with the wrong side, we have been blasting those out of the sky whenever possible. Not all those abducted are returned home. Some are taken to locations such as Dulce Base, put into cages and subjected to gruesome and horrific genetic experimentation. They will spend the rest of their miserable lives in those cages.

"Some others end up being sold off-world on the interstellar slave market. Humans are quite popular as slaves, it turns out.

"The human soul, as it happens, can be accurately described as a form of energy. As such it can be mapped, recorded, duplicated and stored as necessary. It can be removed from the body and inserted back into it…or into another body, depending on what is deemed necessary.

179

"They control your military, your money supply and banks, your food and water, your electrical grid...everything that matters, they control. The bottom line is that you will now do exactly as you are told to do...or you will starve. It does not particularly matter to the aliens which of those options you choose."

I think most readers will certainly now have a better understanding of the problems associated with the idea of full public disclosure. All things considered, it is abundantly clear that there has never been any realistic possibility that full, complete, transparent, unredacted public disclosure would occur. Virtually any of the points listed above would, all by itself, be viewed by the government as enough reason to maintain the policy of official denial and take any discussion of public disclosure completely off the table.

When the tremendous downside to the government which disclosure would immediately cause is considered, there is no President who would voluntarily put that process into motion, if any other choice were available. It makes no difference at all how much the public may think they are entitled to it, or how much pressure they can bring to bear on the matter. No sane, rational national leader with even the slightest trace of an instinct for self-preservation would even consider allowing, much less participating in, something like full public disclosure.

Whether we agree with that difficult and aggravating truth or not, it still represents the bottom line as far as we are concerned. We will be forced to accept that and find some way to live with it for the foreseeable future. For those who have the power to compel or prevent its occurrence, the idea of full public disclosure is a non-starter, something which is—and has always been—completely out of the question.

There is more to it than just the self-interests of those who are directly involved in committing major criminal offenses due to their involvement with alien contact, and we must never lose sight of that fact. An announcement which has the potential to bring down the entire government, and possibly result in the execution of numerous political, intelligence, and military figures, is only one aspect of disclosure.

It also represents the most momentous, profound, paradigm-shattering and revolutionary event in human history, not just for the citizens of the United States, but for all citizens of the world. And once that Pandora's Box is opened, there can be no turning back from it later. Once it has been done, it is done forever and cannot be undone again, ever.

Despite the hopes and belief of many members of the ufology community, it will never take place until or unless the aliens themselves decide to make their presence here known in no uncertain terms.

CHAPTER 20: EXO-POLITICAL REALITIES

In 1626, a man named Peter Minuet convinced a group of Native Americans to sell the land which is now called Manhattan Island for $24 worth of shiny glass beads. The Native Americans, being far less sophisticated in matters of both trade and deception, were convinced to place a high value on the beads, having no idea that they were nothing more than useless trinkets which had basically no intrinsic value. They were convinced they had made an excellent bargain. Manhattan is now the location of the most valuable and expensive real estate in New York City.

"I'd love to change the world,
But I don't know what to do..."
-- Ten Years After

Though it is of course hidden from public view, it is a fact that interstellar diplomacy on the part of the United States has been a fact of life in the world of black operations for the better part of a century. For an outsider, it would appear to be difficult to view U.S. exo-political adventures in a way which is at all complimentary to the government. Indeed, it certainly seems to be the case that the government jumped into the business of interstellar relations feet first, with far too little knowledge regarding who our visitors were, what they represented and what their agenda was. A huge price has been paid for the short-sighted and negligent policies which resulted from what could accurately be described as little more than a fiasco perpetrated by individuals who, despite their power, were not in fact well-qualified to be put in charge of making the decisions they made.

The Best Possible Situation

Presuming that we may at some point be given the opportunity to start our exo-political relationships again, perhaps with a clean slate, it is incumbent upon us to learn all we can form our predecessors and their mistakes to avoid repeating them as we go forward. We must also carefully consider the realities of the situation and give lengthy forethought in terms of what would be the most necessary and beneficial agreements we might someday be called upon to make.

Our first and by far our most urgent concern and responsibility, long before we even begin to consider what form such things as inter-species trade or commerce might take, is to determine our value as a race as seen through the eyes of extra-terrestrial civilizations. We will need to know, for certain, whether they have a compelling reason to want to keep us around and, even more, to do business with us.

We had all better certainly hope that the answer to that question is "yes," because the potential ramifications of any answer but that one are so grim that they do not even bear thinking about.

Provided we survive this first phase and come out of it relatively intact the first order of business, it seems to me, should be to set boundaries on our planetary territory and domain. What this means is that we must lay unquestioned claim as a people to Earth and assert our right to limit and strictly control the numbers and types of extra-terrestrial visitors who travel to our world and the activities they can undertake once they arrive.

That is a rather long way of saying that we first need to establish our borders and assert our right to defend them as necessary, just as any other intelligence civilization could certainly be expected to do were they in our place. Without a consensus regarding our right to stake a claim as the primary owners of Earth and the ability to regulate and restrict the flow of individuals, goods and services according to our best judgment we do not have a diplomatic leg to stand on.

The Difficult Truth: We will find ourselves in a position of permanent weakness compared to virtually all other known extra-terrestrial races whenever we attempt to negotiate any type of diplomatic agreements.

We must always be keenly aware of the fact that as the species with the lowest level of natural intelligence as well as by far the least experience in terms of interstellar political machinations and the intrigue which will almost certainly be inherent to such relationships. That fact will surely constitute a major disadvantage to us and it is quite likely that we will be occupying that position on a permanent basis.

184

The Difficult Truth: We know very little about the races with whom we will be dealing. Most of them, however, are intimately familiar with humanity due to a history of extensive experience and interactions with members of our species.

One aspect of the disadvantage we will probably have to learn to accept is that we do not have the ability in practical terms to infiltrate their forces or their Government. They, on the other hand, have already proven conclusively that they possess such abilities and have an annoying habit of making use of them it whenever it is deemed to suit their purposes. It will likely be highly difficult for us to legitimately verify the compliance to certain agreements on the part of extra-terrestrial civilizations whose main habitations may be beyond our reach and whose technological advantage is of sufficient advancement that it is reasonable to presume it would give them the ability to deceive us or to falsify information which it is not within our power to validate or invalidate.

We must also understand that we are quite limited in terms of coming up with something we can offer a highly-advanced civilization as a form of trade goods that they will find useful and cannot already produce themselves without our assistance. They, on the other hand, will possess any number of technologies and capabilities which we will certainly want to avail ourselves, some of which may even be urgently necessary for us to acquire.

This leaves us in a position in which we may be subjected to radically unfair terms in any potential trade agreements which might be discussed. It also leaves us vulnerable to straight-out blackmail: we could be told that certain necessary technologies or products will be withheld from us until and unless we comply with the demands of an alien civilization which may not have anyone's best interests in mind but their own. I do not mean to imply that this will necessarily be what happens, only that we must keep the possibility in mind as a matter of our own protection.

Humanity would certainly be well-served by choosing our ambassadors and representatives to other intelligent civilizations with great care. Considering what is at stake in our future interactions with these beings and civilizations, it would behoove us to ensure that those who represent us are the most intelligent and well-qualified individuals we can come up with rather than instituting a practice such as the one we currently use when dealing with other earthly nations, when political payoffs and favoritism are often the defining qualifications for ambassadorial status. In a situation where mistakes and misunderstandings could potentially carry with them the gravest of results, those who will represent us must be chosen with great care and only after long deliberation.

Great care must also be taken to ensure that any potential agreements which are made between ourselves and an alien nation are done with the intentions of being beneficial to humanity rather than being designed only to bring benefits to certain selected special interest groups or powerful individuals. There is no room for playing favorites in this sense—interstellar diplomatic relations represent a global paradigm shift of enormous import and its effects cannot be underestimated.

Though we do not know with certainty just when it will occur, at some point it is inevitable that the presence of extra-terrestrials will be made known in no uncertain terms to the citizens of the world...and when that day comes, everything we have ever known will change. The infancy of humanity will end in that instant. Our presumption of isolation as well as our view of our own capabilities will never again be as they are now.

The nature of reality, the limits of possibility and our own place within the universe will be massively re-ordered and it will not be possible to ever return things to the way they used to be. It is no exaggeration to say that entire societies will undergo massive change and restructuring virtually overnight. Our concepts of our own history as well as our faith in our scientific understanding of many things will be shaken to the core.

There will inevitably be many among us who are not capable of adapting to this new paradigm and who will rebel violently against it. Even if the greatest care is taken when the day of disclosure arrives, there are segments of the population which will panic and riot. Damage will be done, and people will die during the transformation from the old ways into the new reality. This is unavoidable in my view. Governments will fall and be replaced by new ones. Whether those new governments will be of the type which will repress their own citizens to an even greater extent than they currently do or whether they will open the way to a future of freedom and prosperity for all is a crucial question which cannot be answered until those changes come about.

The door to the future will be opened wide on that day and our destiny will, with luck, be in our own hands at last. The choices we make then must be carefully considered and made with the long-term welfare of humanity as the over-riding top priority. Humanity's track record has not been a good one, either in our dealings with extraterrestrials or our dealing with our fellow humans. For the sake of all who will come after us, we must do everything within our power to make sure we get it right if the opportunity presents itself. Everything will depend upon our ability to do that.

The True Situation

The most cursory examination of our place within the scheme of galactic intelligent life hits us squarely with something which will be highly difficult for many people to accept.

The Difficult Truth: It certainly appears that we are the least intelligent, least advanced "intelligent" race known. We will be unable to out-think, out-fight, out-strategize or outmaneuver virtually any members of this veritable multitude of races which confronts us.

We appear to be literally surrounded by a multitude of beings which are far more intelligent and far more advanced technologically than we are. The gap between their capabilities and our own is so vast that it is highly unlikely that we will ever be able to close it to a significant degree, much less to "catch up to" or surpass them.

We have only to look at our own history to clearly see the problems faced by a civilization which possesses only a primitive level of technology when it encounters one possessed of a much higher level of scientific understanding and technical capabilities. It is an understatement to observe that the result of such a meeting has never been to the advantage of the less-advanced civilization.

The most common—indeed, the apparently inevitable—result when such a situation comes about is the absorption or the utter destruction of the more primitive group by the civilization which can bring its technological advantages to bear. The Native Americans, to take just one of many examples from our own history, did not determine what the outcome of their interaction with the peoples of

187

Europe would be. They did not set the conditions for the field of battle, they were unable to defend themselves against what turned out to be massively superior weaponry and military tactical capabilities and they did not have the ability to present any type of long-term threat which could serve to stop or even to significantly slow down the expansion of the territory controlled by those of European descent and the diminishment of their own.

When it eventually came to pass that treaties were negotiated between the two sides, the Native Americans had little choice but to let the other side set the terms of the treaty and to take whatever it was that they happened to be offered. They did not have the ability to hold out for more, to make demands of any kind which would dissuade or deter the plans of their opponents or to project any type of military threat which would serve to intimidate the other side. One does not, after all, threaten a group armed with Gatling guns and artillery with bows, arrows and tomahawks. In the end, they had only two choices: they could accept the white man's terms and take what they were offered...or refuse and be exterminated.

That is clearly a position no rational person would ever wish to be forced to "negotiate" from. It gives one side all the power and influence and lets them set the terms of any potential agreement in whatever way they choose. The other side, unable to realistically threaten their opponents and ultimately incapable even of defending their homes against aggression, whether warranted or unwarranted, has no alternative but to capitulate. Their civilization and way of life effectively comes to an end at that point and their future rests entirely in the hands of their enemies.

Few situations in life can be more unpleasant or more inherently threatening than that. It is not an exaggeration to point out that it was quite possible for a single individual—in this case, the President of the United States—to utter a single sentence to his military commanders and the result would have been the complete extermination of the Native Americans, down to the last man. Not a trace of their culture or their peoples would have remained had that been the case and no matter how morally correct or noble the Native Americans may have been they would have been utterly powerless to prevent their own destruction from occurring.

This is just one example out of many which could be examined, all of which have the same basic result: the more advanced and powerful civilization ultimately decides virtually everything up to and including whether the less advanced civilization will even be allowed to survive. The less advanced civilization is always in a no-win position it has no way to escape from. It is entirely at the mercy of its opponents with no way to retaliate against them and no escape possible. Is this not the situation we find ourselves in when faced with literally dozens of extra-terrestrial civilizations, all of which are vastly more advanced than we are and all of which are in practical terms immune to any threats we can bring to bear? In the end, we do not have the ability to threaten their home worlds in any manner whatsoever...yet all they must do at any point is drop a large asteroid into our atmosphere and we would be permanently wiped out.

There has never to my knowledge been an example of a more powerful civilization volunteering to donate its own resources and energy to the cause of assisting a less-powerful civilization to advance technologically and exist independently for the sake of philanthropy or a selfless desire to see others succeed at the expense of their own nation. When such assistance has been offered, it has always been incorporated into a larger long-term plan under which the less-powerful civilization was conquered by or became vassals of the more-powerful nation. The matter of which would eventually dominate the other was never in question—rather, it was a foregone conclusion which the less-powerful group did not have the ability to change.

In every case, the group with more advanced technological capabilities held an overwhelming advantage. It was impossible in practical terms for the other side to do anything in the end but capitulate or be destroyed.

It is in these stark and uncomfortable terms that we must view ourselves when it comes to matters of interplanetary relations with advanced cultures. I see no other realistic way of viewing the situation. Compared to our visitors we are—completely regardless of their intentions or agenda—powerless in practical terms.

The technological advantage they possess is of a far greater magnitude than the one which was present in our examination of the problems faced by Native Americans when the Europeans arrived on their shores. In that case the difference in technology could be measured in terms of somewhere around a couple thousand years. In the case of the intelligent civilizations which clearly already surround us—as well as many, many more that we are currently unaware of, but which surely exist—we are dealing with a technological gap which is truly profound. Some of these extra-terrestrial civilizations have had interplanetary travel since long before homo sapiens even existed.

It is quite likely that some—perhaps even most—of these groups also possessed technology which allowed them to move between dimensions as well for a similar amount of time. The idea of intelligent life existing in other dimensions is something which our mainstream science has not accepted to this very day. Such concepts are considered theoretical possibilities at best and are far more likely to be labeled as pseudo-science and tossed away without being given much thoughtful consideration in scientific circles.

We will be dealing with beings who had the ability to develop technology capable of utilizing in a practical manner concepts which most modern scientists do not even believe to be possible…and the extra-terrestrials have had this technology before our distant ancestors had even come down from the trees, much less come up with the idea of sharpening a stick. That is a technology gap which can potentially be measured in millions of years.

A valid comparison which would illustrate such a gap might be to compare our modern orbital laser weapons and nano-technology to a group of chimpanzees whose proudest achievement was the discovery that they could use a hollow piece of straw to suck ants out of a tree stump. It is not a gap which it is possible for us to narrow in any meaningful way regardless of how much time passes.

In a situation involving military conflict with any of them the best we could realistically hope to achieve would be to delay their victory. We do not have the ability to threaten them in any way which would cause them to modify their plans significantly or cease any theoretical hostilities against us. Basically, we could put up a defense of our world which might possibly serve to slow them down, but which would ultimately prove unsuccessful.

To illustrate it another way: it is believed that two of the extra-terrestrial civilizations with which we have come into contact have been continually at war with each other for over a billion years. You can build as many "advanced" weapons systems as you like. It will not change the fact that if they want us, they will have us and there is nothing we can do about it.

We do not now, nor have we ever controlled the orbital space around our world nor have we ever held air superiority even in the skies over our own country…nor is it likely that we ever will. Without control of our own orbital space we have no realistic prospects of any type of meaningful long-term defense. We are subject to the agenda and desires of whoever it is that happens to control that orbital space and we exist only at their sufferance.

To imagine that the situation is anything but that is nothing more than a case of self-delusion. The future of humanity is surely dependent upon the actions and intentions of various extra-terrestrial nations and groups. There is no valid reason to suppose that all or even most of them would have our best interests as their primary concern, nor is there any evidence that such is the case. Rather, it appears quite clear and beyond debate that some of these extra-terrestrial powers have come here intrusively, acting with agendas which do not revolve around our best interests or welfare and which in some cases are unquestionably hostile, threatening and sometimes lethal.

It seems quite clear, in fact, that several of these groups are contending with each other for influence and control of our world and that our wishes do not enter into their calculations in any significant way. It seems further to be a matter of established fact that human history and human societies have long been influenced and/or controlled by the agenda of non-human entities and that they continue to be now.

When the big picture is truly considered, it seems there can be little doubt that even as you read these words your government and your military is being silently controlled and manipulated by non-human predatory intruders who care not at all about our long-term prosperity. They do not, as best we can tell, recognize any claims to "natural rights" or sovereignty that we may imagine we are somehow entitled to. Our desires have little or nothing to do with their agenda and our abilities can in no way obstruct their plans in a manner they do not have the ability to counteract or overcome.

Whether we want to admit it or whether we choose to spend our lives in denial of the facts, we are already slaves to alien masters and it seems we are destined to remain so for the foreseeable future.

Without engaging in any moral or spiritual judgments, it is clearly the fact of the matter that we have been and are currently being manipulated and controlled by beings of vastly superior technological capability and experience who have no reason to care whether we like it or not. We will find ourselves in no position to make any significant demands of our extra-terrestrial visitors, regardless of what their intentions may be, nor are we likely to have the ability to prevent them from doing anything they choose to do in the end.

CHAPTER 21: HUMAN RESOURCES

For in much wisdom is much grief:
and he that increaseth knowledge increaseth sorrow."
Ecclesiastes 1:18

"And our time is flyin'
See the candle burning low
Is the new world rising
From the shambles of the old?
If we could just join hands
If we could just join hands..."
--Led Zeppelin

Imagine, if you will, a galaxy which, rather than being sparsely-populated and eternally distant, is home to a wide array of intelligent entities or beings, and a wide array of cultures and civilizations. I am not talking about a dozen or so different alien races, I am talking about hundreds of groups of all possible types, maybe thousands.

It's a big galaxy, after all, and I submit that we have been shown in no uncertain terms that we are late to the party and several dollars short on the admission fee.

I know the argument which states, quite correctly, that one cannot judge an entire alien race based only some of its members. Neither is it fair or right to judge an individual by a stereotype idea of its race.

Quite right.

But here's the thing. I don't have the option of getting to know every non-human entity that's out there on an individual basis. And if I did, I'm sure I wouldn't want to. From the looks of it, there are more than plenty of things out there in the reaches of space and time that I want to keep my distance from and do my very best not to irritate too much.

And how much, I ask you, is "too much" to an alien I know almost nothing about? Do you see the problem?

I am writing this book not because I want to, but because I feel I must. There will be those who don't agree with me about certain things among my readers, I know that in advance.

Some of them will surely be proven right. That is fine with me, if we end up with a better answer than we had before.

There will also be those who don't believe a word I say about any of this. That's alright, too. I am just a writer, doing my best to make my way through a godawful maze as best I can. If you tell me you don't believe a word I say, I'll tell you that you have every right to not believe it.

But if you'll follow me on this journey, through its many twists and turns, across vast chasms of thought, spread out across so many seemingly-unrelated topics, I can—with the kind permission of those who watch—share with you some things I have learned.

I am not doing it with the intention of challenging or upsetting anyone. I am just a writer, just one person. I have no illusions about being able to challenge any of you, I assure you. I recognize that I am publishing this manuscript only because I am being allowed to do so by people far more powerful than myself, and I thank them for having thus far allowed me to express myself freely.

It is, as you can imagine, often difficult to know with certainty where one should draw the line. It is a balance I try my best to find. What more can I say than that?

I am not here to offend anyone, nor am I here to waste anyone's time. I am here because I believe that, full of faults as we all are, the best examples of humanity are rare, and beautiful, and precious, and worth preserving if it can possibly be done. I am not trying to bother anyone. I just want to write my books, if nobody minds too much.

So, listen, brothers and sisters: a galaxy loaded with life! A galaxy where interstellar travel was common among—let's use a number I suspect is a very low estimate—100 different civilizations, or 100 distinct alien races.

For untold thousands of years—millions, in some cases—these groups have known and interacted with each other. There has been war, there has been aggression, there has been empire, there has been spiritual ascension, there has been all of this and many things more. Some of these alien races are very, very different than we are. Some others appear to be quite like us, in both temperament and physical appearance.

These alien races, and the many subgroups they could possibly form, are no strangers to each other. They have interacted in some ways with each other for uncounted thousands of years. Many, probably most, of these races surely have official treaty status between themselves and the others.

It is not unreasonable to presume that a discussion about their actions regarding Earth and humanity might occur, and that there are very likely stipulations contained in treaties between alien races themselves which forbid or prohibit direct intervention or interference in this matter. There may very well be aliens who would wish to assist us but are prevented from doing so due to the terms of a treaty they had signed with one of their long-term peers.

It is interesting and fun to let the mind run free and to let the conversation flow with the stream of consciousness—one never knows what curves it will take, or how it will end! But I was speaking here of a galaxy filled with all manner of life, some of it millions of years older than us, and almost all of it far more intelligent than we can credibly claim to be.

Imagine a large city, such as New York City. Now imagine a runty little trailer part in the outskirts. That's our home. Come on in. Have a seat.

So, there's a galaxy which contains an unknown—but sizeable number—of what we would consider to be "alien races", for lack of a better term. Whether some of them do not take physical forms is immaterial for the moment. The galaxy is a very old place, and we have apparently come along quite late in the game, compared to most of the others we know of. We are the least advanced, least intelligent, least culturally significant kids on the block. The other kids all have lots of nice toys, we make do with a microwave oven and a cell phone.

No, don't even get that idea in your head: there is no way any of those other kids are going to trade us their interstellar engine design for some cell phones and a couple cases of Etch-A-Sketch. *We* are the stupid ones, not *them*—remember?

These other races, civilizations, alliances, sub-groups...whatever...each have their own set of resources, goals, agendas, beliefs, social structures...everything. Now imagine all of them having diplomatic relations and probably treaty agreements with most or all the others. There will be wars, alliances, trading partnerships, black markets, piracy, slave trading and every other thing that can be expected to exist in such a multi-player galactic game. We should expect relationships between these non-human groups to be widely-varied but commonplace.

This little setup alone gives one much to contemplate, and many possibilities to consider. Perhaps we shall do some of that another time. But for now, I wish to turn your attention to currency. Galactic currency. What do they trade for, and what do they accept in trade?

We can presume that they surely trade many things, between them. We should also presume that many things might be accepted in trade, depending upon the parties involved and the circumstances. We can, and probably should, presume for the moment at least that trade exists in the currencies of gold, slaves, precious minerals, foodstuffs, exotic plants & animals, armaments, proprietary technologies, information, and DNA.

Now, which of those things can be gotten from Earth, if one has the power to get it?

Probably all of them, with the exceptions of armaments and proprietary technologies.

Why do aliens come here in the first place? Probably to get some of those things for themselves, and then either return home with it or trade it for something else.

We should expect that some aliens come here seeking gold or precious metals, some come to collect plants and animals for assorted reasons, some come to abduct people and sell them into slavery, some come to stock up on food or fuel, and some come to collect, store and modify the DNA of a multitude of life forms to be found here, but most importantly that of humans.

Imagine a culture, a very ancient one, let's say it has had interstellar travel for a hundred million years. A hundred million years is a long time. A society can change a lot in that amount of time, and it can learn a lot.

Imagine that this society had learned a brutal truth during its many millions of years of existence. It had learned that the most valuable thing we will ever possess, is our DNA. There is no physical object which cannot be replaced, no accomplishment that will be remembered for a million years, none in this world. None, that is, but one: DNA.

In a society such as this, DNA could be expected to become currency. We ourselves are made of DNA—meaning that our DNA would logically have a certain value to the ancient society in question, a value which would presumably be dependent on the origin and characteristics of the DNA in question. Maybe one of us has DNA which would equate to a $100 bill in their society, maybe someone else's DNA would be worth only the equivalent of a nickel.

Some people, according to this model, would be abducted and never returned to their homes, sold as slaves to others. Whether such sales can take place legally or are accomplished on what amounts to a "black market" is uncertain but makes no real difference in terms of our discussion at present.

In other words, we must presume that humans can be utilized as resources by some of these alien groups, and possibly by all of them. If that is so, then we must also presume that humans *are* being utilized as resources by at least some of the non-human entities in our vicinity.

If we accept that these things are likely to be true, we can then understand why some of these aliens appear to think of us in much the same way that we think of livestock: we are resources, to be harvested as needed. Our planet, then, can basically be conceived of as a big DNA farm, which can also be used as a source of gold, copper, lodestone, water and other resources.

In a galaxy filled with competing interests and diverse agendas, we are the youngest and least-intelligent known race. We are also certain to be the least capable of defending ourselves against attack by hostile alien forces—not just military attack, but attack by *any* method.

In a galaxy which contains an abundance of alien races and civilizations which differ greatly from each other, we can expect the notion that "might makes right" will almost inevitably be employed among at least some of them. This leads to the troubling idea that aliens could well be utilizing us as unwilling resources, regardless of any political agreements which may be in place between themselves and others, or which may apply in other regions of space.

If anybody is looking for a weakling to pick on, it is obvious that we represent by far the weakest weaklings of all. We are the least intelligent, the easiest to trick and deceive, the least capable militarily and—perhaps best of all, from their point of view—the least likely to start a war we could only lose over the abduction of a few million citizens—a statistically insignificant portion of the total population.

After all, what are we really going to do about a group of hostile aliens which engages in abducting our people, but returns most of them home again afterward, and only sells a relatively small number of them off to the slave market? Especially if—and this is entirely possible—those who are chosen to be sold as slaves are people who have been determined to be genetically defective in some way? This would make them (by alien standards, at least) the least valuable and most expendable individuals to be found here.

What are we realistically going to do about a situation like that? Declare war on a highly-advanced, super-intelligent alien civilization, one which is surely capable of decimating our planet at will, when we are unable to even strike any targets of value other than some of their spacecraft which happen to be picked up on radar?

No military commander in their right mind would see this as a sensible proposition. One does not start wars one cannot win. Instead, one plays for time as best one can, and attempts to develop some type of weapon which might be useful against them while it is possible to do so.

It is a historical fact that, in extreme circumstances, the only reasonable option is to surrender without a fight rather than face the certain defeat and destruction that would result from an all-out war. It is, after all, almost always better to surrender than to cease to exist.

Tough times require difficult choices, and it must be acknowledged that this is a difficult choice which could very easily have been in play throughout the modern age of UFO sightings.

It has been said by many that the Greada Treaty, signed in the late 1950's by President Eisenhower, was for all practical purposes a document which laid out the terms of America's surrender to unfriendly alien forces. From what I have been told about the document, it is difficult to disagree with that assessment.

We were unable to fight them with any reasonable hope of success, so we sued for peace, gave them everything they wanted, and took whatever we could get in exchange for it. That is, it seems to me, a fair and accurate description of the provisions contained within that treaty. The specifics of that treaty are discussed within the pages of the previous volume (*"Alien Contact: The Difficult Truth"*).

Now that we are being real about things, let's continue to do so a bit longer. Let's run all this another step or two forward and then take another look at things and see what we find.

In hindsight, it seems abundantly clear that, upon seeing they were faced with an opponent they could not prevail against, those who had the ability to do so attempted to make the best possible deal for themselves. It is also clear that they did this without regard to the welfare or long-term survival of the members of the society they were sworn to protect.

In any confrontation against a highly-advanced alien power, a lot of human DNA is almost certainly going to be lost in the process. Just as I predicted earlier in the book, they attempted to somehow gain an advantage for their own genetic line.

Ultimately, when facing a possible existential threat, the DNA of others would not have been their concern. It is the preservation and safety of their own specific DNA which would have naturally taken priority.

Remember: genetic warfare is ceaseless and without mercy. It takes no prisoners, and its rule over history is absolute. In a contest between your DNA and their own, someone else will always choose their own—meaning they will choose to sacrifice yours, in favor of theirs.

In the end, this is what has been done by those at the top of the food chain when it comes to negotiating agreements with alien civilizations and controlling access to the hyper-advanced technologies we gained as a result. It soon became clear that we had allied ourselves with a group of inherently hostile alien powers, and that their capabilities were vastly greater than we had imagined. It was learned that their long-term intentions regarding Earth and its human inhabitants were anything but friendly, and that they viewed us as little more than ants in an ant farm.

At that point, the priorities of those in charge of the black ops segment of the military, our intelligence agencies, our most powerful corporations and the financial elite of the world underwent a radical change. The interests of the average citizens—the very people most of them had been charged with protecting—fell by the wayside, and their personal interests became the primary focus of their lives.

They then proceeded, in some cases, to negotiate their own individual agreements with these alien races, separate and apart from any agreements which had been made with the American government. This eventually resulted in what are known as "breakaway civilizations", groups of humans who live off-world, have the advantages of unimaginably advanced technologies of alien origin, and operate completely independently of the rest of humanity.

Our tax dollars are used to provide them with massive amounts of funding, all of which is hidden within the secret budget, but we derive no benefits whatsoever from this. This is a topic which will be discussed in greater detail in a future volume. The secret underground colonies we have established on Mars are an example of a breakaway civilization—but they are far from the only one!

Those who have remained on Earth have maneuvered themselves into a position whereby they and their descendants will be assured of having access to these same highly-advanced alien technologies, which will be denied to others. They are also given exclusive access to, and preferential treatment by the aliens themselves, as well as a monopoly on both the ability to communicate with them and the ability to be fully-informed of everything we know about them.

If you were in their position, would you do the same? If you chose not to, wouldn't you be responsible for placing your genetic heritage—your DNA, to put it another way—at a permanent and insurmountable disadvantage compared to those who did?

What if you came to believe that inter-dimensional alien beings, along with their half-human hybrids, had already effectively taken control of the world's governments, financial institutions and military-industrial complexes? And that they considered the rest of humanity to be expendable, and intended to either eliminate or enslave all but a few of them? Would the genetic imperative to safeguard your own particular strain of DNA—an instinct so overpowering that it, more than any other factor, has determined the course of history—allow you any other choice but to protect it?

As can be seen even from this greatly simplified chain of logic, the subject of alien contact is neither straightforward or cut-and-dried.

Many possibilities exist, and many more surely exist which we are not even aware of.

At best, humans as a race are likely to be treated by aliens in much the same manner that ufologists have been traditionally treated by our own government: we may be allowed to see only a tiny part of the whole, denied knowledge of things which could easily result in an entirely different view of things, were they to be revealed.

At the very least, it seems fantastical to me that anyone should make the presumption that we humans are owed some type of assistance, tolerance, advantage or anything at all by *any* of the veritable multitude of other races and powers which surround us.

The alien powers can easily look upon our supposedly "advanced" civilization and see that we blithely allow our own crippled veterans to take shelter on park benches and sidewalks during the freezing, heartless winters. After doing so, they might quite properly wonder what benefit or advantage to them would be gained by assisting us, at their own expense…and come up with an answer of "none".

What kind of price might they potentially incur by deciding to try to assist us anyway? We have no way of knowing the answer to this, but I can think of several possible scenarios which could easily cause the idea of assisting us to be unacceptably dangerous from their point of view.

All things considered, it might well be the case that we are effectively left alone here, to solve our own problems or *not* to solve them, and to live with the consequences of whichever it may be. Is it sensible to assume that this is not the case, without first being provided with some sort of compelling proof that serves to demonstrate that our assumption is true? Or is it, rather, little more than wishful thinking and self-delusion to imagine that we somehow "deserve" special consideration and outside assistance, while having little to offer in return which cannot already be freely taken by any who desire it?

If we act according to the presumption that assistance from our alien brothers will be forthcoming, and no such assistance materializes, will we not have rendered ourselves helpless and vulnerable by waiting for it, rather than scrambling as quickly as we can to tend to our own defenses? If, on the other hand, we presume that we are on our own, and act accordingly, but alien assistance arrives unexpectedly, will we not then be in a better position to receive and utilize it than we would have otherwise been?

The best—and, to my mind, the only—logical solution to this conundrum is to make use of the adage "expect the best but prepare for the worst!". Doing anything less than this is, it seems to me, not only the height of foolishness and irresponsibility, it could very well be a mistake which proves fatal.

In this game, there will be no second chances. What that means, in practical terms, is that we do not have the luxury of making the wrong decision and hoping for a chance to correct it at some point in the future. There is no reason whatsoever to presume that we will be given such a future chance and—as we have seen—there is no reason to think that we would be considered deserving of one by the alien powers which surround us.

We dare not make mistakes. If we do, they could well prove to be the costliest mistakes imaginable and we will have no chance to correct them later. We have got to get this thing right the first time around. This leads to an important question. It is one I have never heard discussed in ufology circles, but which absolutely should be. It is this: how can we be certain of making the correct decisions when critical information is withheld from us and we are unable to assess the situation accurately?

The short answer is, "we can't". We are not, however, charged with making those decisions, nor would our input be given any weight if those who are were to ask us.

The question, however, does not come to a rest there. What if those who are charged with making those tough decisions find themselves to be in the same position we are in, regarding the aliens? What if they, too, are being forced to decide and act based on an insufficient amount of information, or disinformation the aliens have convinced them represents the truth?

Could they then be expected to make the right choices?

Is there any reason to assume that they are *not* in this position, when it comes to dealing with the alien visitors? If you know of one, please contact me and let me know what it is, because I do *not* know of such a reason, and I have thought long and hard about this matter.

If we truly wish to arrive at a place where we can at last fit the pieces of the puzzle together, we must begin by acknowledging our status on the ladder of power.

We are, and always have been, human resources which are controlled—and, ultimately, harvested—by alien masters. We are pawns in a game we cannot even begin to understand. Rather than being marched down the board and promoted, we are far more likely to remain pawns forever.

That is not an encouraging idea to think about. There is nothing about it which can be considered "happy", "fun" or "inspiring". That is unfortunate, to say the least…but I am not responsible for the answers.

I ask the right questions, to the best of my ability—that is the contribution I try to make within the text of these manuscripts. Then, like all of you, I wait for the answers to arrive. I hope that, when they do, they represent something positive.

Sometimes they do. Often, they do not.

Whatever the case turns out to be, I pass them along to the reader in the same condition I find them, as data to be considered. It is not usually a whole lot of fun. But it is what I seem to be here to do, so I do it to the best of my ability and hope it might somehow be useful to others.

With luck, it may end up reaching enough people to make a difference, even if only a small one. If so, then I have done my job and this book is a success.

With all of that said, and with the hope that each of you will choose to return for our next round of discussions, I shall proceed to the concluding section of this manuscript. It is something rather special, and I have included it as a personal gift to those who have been kind enough to purchase this book and made their way through it.

Thank you for allowing me the opportunity to share this information with you, whom I consider to be my brothers and sisters. In the final analysis, you represent the few, not the many. Most will never own a copy of this book, nor will they make any attempt to read and understand its contents.

Those who do will hopefully possess clarity of thought, an elevated level of situational awareness, intellectual acumen and a great deal of courage as well. It is for you that I created these volumes, and it is to you that I entrust them with complete confidence and my highest regards. I feel that you deserve to have material such as this made available to you, and I trust that you will make use of it in ways which are proper, fitting and necessary.

As an author, this is all that I can hope for and all that I can reasonably expect. It is, in my judgment, enough. I will be satisfied and content in the knowledge that I have cut no corners and delivered a manuscript which truly represents the best work I am capable of.

All that remains to be done is to release this creation, this product of my blood, sweat and tears, into the universe, confident that it will eventually make its way to its intended audience and be assigned whatever value it deserves. This I shall gladly do.

So mote it be!

CHAPTER 22: SHINE!

"Whatsoever thy hand findeth to do, do it with thy might,
for there is no work, nor device, nor knowledge, nor wisdom
in the grave, whither thou goest."
-- Ecclesiastes 9:10

"Mama always told me not to look into the sights of the sun
Whoa, but mama, that's where the fun is!
I was blinded by the light,
Blinded by the light..."
--Bruce Springsteen

In an earlier chapter, I made a point of chastising the group of people who claim to be able to channel messages from alien beings. I once believed, because of watching a lady called J.Z. Knight on television many years ago, that all channelers were con artists.

She claimed to channel the spirit of an ancient Roman warrior called Ramtha. A member of the audience said "If you lived in ancient Rome, then you must know Latin. Speak some Latin for us."

J.Z. Knight, in the alleged persona of "Ramtha", stuttered and mumbled for a moment, then triumphantly practically shouted "Veni, vidi, vici! There! Are you satisfied now? Have I proven myself?" I consider it to be the most transparent and blatant example of fraud I have ever seen, to this very day. Unless someone attempts so enter a Sea Monkey in the Kentucky Derby, I doubt my opinion about that will change.

One day, around fifteen years ago, though, I came across a method for channeling and decided to give it a try, not expecting that anything would result from it. What happened would eventually change my opinion about the phenomenon of channeling in general.

All I did was silence the voice that we all have running through our head all the time, and then listen. Much to my amazement, almost as soon as I did this, I heard another voice begin to speak!

It was not a voice I was familiar with, or one I had ever heard before. I would compare it to listening to a university professor delivering a lecture to a class of students. I did not go into a trance, or a state of altered reality, in any way—I remained completely lucid the whole time. Neither did I take on the "personality" of another being—it was just like hearing a highly intelligent professor speaking to a classroom of one.

As it happened, I was sitting in front of my computer at the time I began hearing this. I am a very fast typist, and I started typing as quickly as I could, recording word-for-word the things I was hearing. It continued for a half hour or so. Then, when whoever was responsible for the voice had said all it wanted to say, it stopped.

The words on the page that I had recorded spoke of things which were somewhat unfamiliar to me, and at times even over my heard—I didn't have enough background in the topic to understand them clearly. Even so, they were quite remarkable and surprising to me.

I refused to believe that what had just occurred was an example of channeling, however. I presumed it must have been my subconscious mind speaking to me somehow, though how it could speak about a subject I was unfamiliar with I couldn't explain.

A few days later, I decided to try the experiment again. As before, the voice began delivering a lecture. It did not pause for questions, it did not allow me to choose the topic of the day, and it continued until it had said all it wanted to say, then stopped. As before, I put it off to nothing more than an odd trick of the mind, rather than an example of channeling.

Over the next couple of years, however, I continued to receive this type of transmission whenever I attempted to channel something. The topics the messages addressed were highly varied—the best way I know of to describe them is that it seemed to me that it was dictating sort of an instruction manual for the soul, if that makes sense. Each time, I faithfully typed out what I was hearing, word for word. Each time, I refused to believe I was really channeling anything.

Twice during this time, I heard—and typed out—what appeared to be nonsense phrases or gibberish. They were not wording I was familiar with, nor were they in a language I recognized. I ignored them, setting them aside and continuing to receive and record additional messages.

I would note that I did not really enjoy the process of doing this very much. It took a lot of energy, and by the time a session ended I was usually worn out. When a "lesson", as I came to think of them, was nearing an end, I would often hear the voice say something along the lines of "You are tired now. That is enough for tonight. Go and rest yourself. Return when you are refreshed, and we will continue".

As with the first message, many of the subsequent messages dealt with topics I had little familiarity with and did so in considerable depth and with authority. It spoke of the astral plane, of psychic abilities, of possessing the ability to command and control angels and demons. It talked about communication, and planes of energy, and personal morality and conduct. It spoke about the nature of the soul, and the timeless nature of consciousness. The subject changed each time and was beyond my ability to alter or request.

Eventually, I found that I had collected a group of messages which, when assembled in order, basically formed a book of around a hundred pages or so. I would describe it as a combination of an instruction manual for the soul, a guide to moral behavior, and an instruction manual for certain aspects of astral travel and ceremonial magic.

It contained nothing harmful or evil, nor did it attempt to control my actions in any manner. It was, as best I could determine, a textbook intended for me to read, study, and learn what I could from. I had no idea why it happened to be given to me specifically, other than the fact that the voice had mentioned several times that it was given to me because I had the ability to receive it. I had the ability to hear this voice, when others apparently did not.

The voice also mentioned several times that, although much of the material it spoke of was over my head, "there will be others who have the ability to understand it, and it is they whom it is intended for. Find them, and they will explain it to you."

During the process of all this, I suffered the failure of both my hard drives within 24 hours of each other. Though much of what I had recorded had by then been transferred to disk, I unfortunately lost some of the material to this hard drive failure, including one of the phrases which I had taken to be nonsense or gibberish.

Before this happened, though, and about six months after I had received those nonsense phrases, I happened to look at them again. It occurred to me that one of them looked as though it could possibly be Latin, and that I could check this by submitting the words to one of the translation dictionaries which were available on the internet. I decided that it couldn't hurt anything to give it a try, and I did just that.

I do not speak more than a few simple words in Latin. It is a language I have never needed to use in my everyday life, and one I have never had any particular interest in bothering to learn. That is, in fact, the case with all non-English languages: I made a conscious decision not to study or learn them, and I was successful in not doing so.

When I was a boy of fifteen it occurred to me that, for me personally, it would be far more useful and valuable to invest the same amount of time many people spend learning a foreign language into becoming twice as good at English as I was before. I have never had reason to regret that choice. It has proven repeatedly to have paid off for me and I highly recommend it, especially for those who aspire to become professional authors.

If asked how much Latin—or, for that matter, any other foreign language—I know, the most honest and accurate answer I can give is "as little as possible".

The only exception to this is a pidgin language known as "horse Latin", which I learned from the late Richard Bach, author of "Jonathan Livingston Seagull" and other books. Though we are all familiar with "pig Latin", most of us have never even heard of "horse Latin", and I was no exception.

To speak "horse Latin", one simply adds an "iv" or an "ivi" into the center of ordinary English words. Whiven spivoken, ivit sivounds livike thivis.

Though it has virtually no practical value for the most part, I took a liking to it because it tends to completely baffle illegal aliens and make them half-crazy, and I get a sadistic sort of satisfaction from being able to do that on occasion. That, however, is pretty much as far as my knowledge of fivoreign livanguages givoes.

At any rate, I eventually decided to look the words up in some Latin-English dictionaries online, just on the off-chance that it might turn out to be Latin of some kind. When I did, I got a rather large shock.

It was Latin alright. Not only that, it was a phrase in perfect, flawless Latin. I compared the words between probably a half-dozen different versions of Latin-English dictionaries which were to be found on the internet, and every one of them agreed.

When what I had taken to be a nonsense phrase was translated into English, it read:

"A sign (or seal) of authority, a little book of truth."

:-O

I am not sure just how to argue with that.

In the end I—the person who was probably least willing of all to classify these messages as something which is an example of legitimate "channeling"—found that I had little choice other than to do so. I am unable to find any other convincing way to explain them.

If the reader feels it best to ascribe what follows to either a trick of my subconscious mind, or to believe that I simply shifted from one writing style to another and invented it myself, this is something that I certainly encourage. I would like nothing better than to be given full credit for writing it myself. If you want to give it to me, please feel free to do so! *

In my opinion, however, that is not the case. I think I am a reasonably good writer, but I do not believe I can create something like what you are about to read on my own. I wish that I could, but I have never done so before and it is a style of writing which I do not recognize as my own.

What the source or origin of the transmission may have been, I am unable to say with certainty and I prefer not to speculate about it. I have an opinion, but it is little more than an educated guess and I see no reason to include it here. The reader is free to believe whatever they like about it.

The material contained in the message you are about to read does not deal directly with the topic of either UFO's or alien contact. Rather than the often dark and gloomy picture I paint regarding aliens in general, its tone is highly positive and inspirational.

It strikes me that there is, perhaps, a lesson to be learned from its tone, one which tells us that despite the dangers and risks associated

---

* You may even wish to follow it up by writing me letters telling me how great I am. If so, they would be the first I ever received, and I will surely have them framed for display on my wall.

with alien contact, we still retain the ability to control our actions and, for the most part, choose our own destinies.

Surrounded by alien entities we certainly are, yet the way we choose to act and to go about our daily business remains ours to determine. With this fact in mind, there is much we can achieve during our lives irrespective of the presence or absence of alien beings on our world.

I have included a single example of the material I channeled here, for your consideration. Though it was a message sent directly to me, I believe most of it can be useful to others as well and can be incorporated into their lives if they should choose to do so.

I chose to include it here both as a way of ending this manuscript in a positive, inspirational way and as my personal gift to all those who have been so supportive of both myself and my work during the past several years.

It is to you, kind friends, that I dedicate this next section. I hope that you will find it to be as useful to your own lives, as it has proven to be to mine. Once again, this is something I received about fifteen years ago at the time I am writing these words.

Here, then, is the message I would like to share with you. Not a word of it has been altered, nor a word added. It is the same now as it was when I initially received it.

"To judge the value or worth of a soul by the color of the skin its body wears, or by the place or circumstances of its birth, is the mark of a fool. Wisdom, virtue and knowledge can be found beneath skin of all colors and resides in the hearts of people from all places.

"The same can be said of greed, evil, selfishness, cruelty and works of dishonesty and ruin. Judge not a soul by the color of its skin or the condition of its apparel, lest you do that soul a disservice and reveal thy own short-sighted and narrow vision.

"If you must judge the value of a soul, then let it be judged by its works and its principles. Let it be judged by its compassion, its kindness to its neighbors and its charity toward those who are less fortunate. By these things is a man truly known, and it is by these things that his destiny shall ultimately be written.

"Never forget that thou are not a body which contains a soul, but rather a soul which has incarnated into a temporary body for a time. It is a certainty that, in the fullness of time, the body you now inhabit will die. It will return to the dust from which it was formed, and yet your consciousness, that which you deem to be your immortal soul, will carry on in another form.

"That form will be determined by the quality of the life you have lived and may not be known or determined by any man in advance. Time will pass, and the form will be revealed in its proper time and place—this is as it should be.

"Act, then, with this always in mind, that you are not merely a body with a soul residing inside, but a timeless and immortal consciousness which will and must be held responsible for every action it takes, and every action it fails to take. Think of the long term, not the short, if you are wise!

"When given a choice between doing what is right and doing what is easy, always choose to do what is right. Do this and you will gain the respect of your fellows, and acclaim which you will rightly deserve.

"You do not have the right to interfere with the Will of another, provided it is doing you no harm. Each soul is on a journey of thousands of lifetimes, and its path and destination are unknown to you. Nor are they a matter of direct concern to you, since it is your primary responsibility during this life to improve the quality of thy own soul and purify it to the best of your ability.

"You are not responsible for the soul of another. Tend to your own instead! Do not force your views on others. Do not presume your way of thinking is the only correct way. Never allow yourself to believe that you have all the answers, or do not need any further improvement. Any who believe such arrogant and foolish things will surely be mistaken.

"Do not engage in gossip. Do not associate with the scurrilous or the criminal. If you do, your reputation will suffer because of it, and this is a price you cannot afford to pay.

"Do not associate with those who are unlucky, for luck is contagious and their bad luck can be transferred to you. Do not associate with those who are negative, for like attracts like and their negativity will surely draw bad luck.

"If you are nearby when that bad luck arrives, it will be transferred upon you simply from being in the wrong place at the wrong time. Do not allow this to happen. Avoid people such as this.

"The world you see around you is an illusion. Never forget this fact. The truth is hidden by many layers of deception, and many appearances which do not reflect reality.

"The real truths are for the few, not the many, and such has it always been. The common man will believe that what he sees, and thinks he understands, is the truth of the matter. The wise will understand that what is seen is but a veil, an appearance which covers deeper and more profound truths which are beyond the understanding of the common man. They shall remain forever hidden, and beyond his grasp.

"If you are one of the few, you will know it. If you are one of the wise, you will feel it within you and you will know. For you, the truths are there to be found and revealed, if you are willing to expend the time and effort required to locate, experience and understand them.

"Do not attempt to teach commoners the secrets which are meant to be reserved for the few. They are not capable either of truly understanding them, or of utilizing them in the manner for which they are intended. Only harm and chaos will result from mixing the common with the uncommon—do not attempt it.

"Each man has his place, and each his level—this is the way of things. Thus, it has always been, and thus shall it ever be.

"Understand that you exist in a universe of balance—light and darkness are finely balanced, in a way which is and shall remain beyond your power of understanding. Neither can, in the end, triumph over the other, nor should it be that they could.

"Darkness can never triumph, for in Darkness is death, and death is no triumph. Neither can Light exist without the darkness, for without it Light would lose its value and become ordinary rather than something to be aspired to.

"A universal triumph should not be what you seek. Rather, seek to allow the Light to triumph within yourself, to radiate from within you and illuminate that which is without. This is the purpose of the soul-journey, and the thing that all men must someday, and in their own time, learn.

"The greater universe itself is far beyond your ability to command or manipulate. The inner universe, however, is entirely dependent upon your thoughts and intentions, your words and deeds and actions. These things, which are within your control, will in the end work to shape the universe as you are able to know it. They will determine the quality of your journey, and the fate of the soul which you have been gifted with.

"Shine! Shine as a light in the darkness, when you have learned sufficiently of the Light to do so! Illuminate the path, that others may see and understand the things you have learned, so to ease their journey and assist in their ascendance as you are able to.

"Shine! Do not be a bringer of darkness, but instead become a beacon of light to those around you! This is one of the great secrets which will serve to divide those who truly understand from those who do not.

"Never settle for Darkness, when Light can be had instead! Never shine with a half-light—shine always in the fullness of your powers! This is the reason they were given to you, and the use for which they are intended.

"There is wisdom here, for those who can understand and make use of it. There is a path to success and victory here, for those who can find it. Again, I tell you, and listen when I do: Shine!

"The soul is a wheel, yes, a wheel, spinning in eternity and caught by the spokes of time, allowed to dance for a little while, then jerked sharply back into line according to the Will of the Masters. Wheels, wheels, all of you, now and forever, for that is the Plan.

"Dance, little wheel! Dance, while the opportunity exists! Dance for us now, little wheel of dust, and feel your true power come forth! That is all we shall ever ask of you, but this you must do!

"Dance, I say! Dance and shine! Then the powers which are within you will manifest in a glorious display of beauty, which will illuminate your path and please those who have set you upon it!"

Readers can find updates, additional photographs, news items, links and other useful information by following this link: www.facebook.com/OfficialDerekTyler/

Class dismissed.

CPSIA information can be obtained
at www.ICGtesting.com
Printed in the USA
LVHW041618250419
615555LV00001B/142

9 780692 948033